WATER SUPPLY ENGINEERING DESIGN

WATER SUPPLY
ENGINEERING DESIGN

M. Anis Al-Layla
Dean, College of Engineering
University of Mosul
IRAQ

Shamim Ahmad
Professor of Civil and Sanitary Engineering
College of Engineering
University of Mosul
IRAQ

and

E. Joe Middlebrooks
Dean, College of Engineering
Utah State University
U.S.A.

ANN ARBOR SCIENCE
PUBLISHERS INC
P.O. BOX 1425 • ANN ARBOR, MICH. 48106

Preface

This book is intended to serve as a textbook in water supply engineering for use in developing nations. It is hoped that this book will also prove useful to the developed nations and the practicing engineer.

Suitable examples are solved to illustrate principles wherever possible. In general SI units have been used, but MKS units are also given and occasionally FPS units are indicated. A table of conversion factors is presented in the Appendix.

The authors acknowledge with thanks the constant encouragement and help provided by their wives and Dr. Mohamad Al-Mashat, President of the University of Mosul, during the preparation of the book. The authors are also thankful to B. Koprulu for typing the manuscript and to Barbara South for preparing the final manuscript. The editorial assistance of Donna H. Falkenborg and Charlotte H. Middlebrooks are gratefully acknowledged.

The authors are particularly indebted to the following firms and individuals who were so kind to grant permission to reproduce drawings and photographs from their books and advertisements: Johns-Manville, Denver, Colorado, USA; Envirotech, Salt Lake City, Utah, USA; American Society of Civil Engineers, New York, N.Y., USA; McGraw-Hill Book Company, New York, N.Y., USA; R. Oldenbourg Verlag GmbH, München, West Germany; BSB B.G. Teubner Verlagsgesellschaft, Leipzig, West Germany; World Health Organization, Geneva, Switzerland; John Wiley and Sons, New York, N.Y., USA; Linvil G. Rich, Clemson, South Carolina, USA; Pitman Publishing Limited, London, England; Infilco Degremont, Inc., Richmond, Virginia, USA; Allis-Chalmers, Norwood, Ohio, USA; Dorr-Oliver, Inc., Stamford, Connecticut, USA; Envirex, Rexnord Company, Waukesha, Wisconsin, USA; Environmental Elements Corporation, Baltimore, Maryland, USA.

Anis Al-Layla, Shamim Ahmad, and
E. Joe Middlebrooks

Table of Contents

1. Introduction

Water is vital for man's existence; without water there would be no life on earth. The body of a human being consists of 65 percent water. Apart from the day to day requirements, water is needed for irrigation, power generation, recreation, industrial production, and receiving wastewater.

There is an enormous amount of water on our planet, approximately 1.4×10^9 cubic kilometers in the form of oceans, seas, rivers, lakes, ice, etc. But only 3 percent of the total quantity of water on the earth is in the form of fresh water available in rivers, lakes, and groundwater.[1] Fresh water is limited, but the requirements for fresh water are ever on the increase, due to the increase of population and industrialization.

According to the United Nations Population Reference Bureau, the world population near the birth of Christ was about 300 million. In about 1680 it was 600 million. By 1850 it grew to about 1200 million, by 1945 to about 2400 million and by 1969 to about 3500 million people. It took 1680 years for the population to double after the birth of Christ. After that it doubled in only 170 years, and then again in 95 years. It may take only 45 years to double the 1945 population, i.e. by the year 1990 the world population may reach 4800 million and by the end of the century 6000 million. The result of all this is obvious; the total water requirement is ever on the increase, and the per capita water consumption is also on the increase.

During the sixties the water consumption in Germany for domestic, public, and commercial purposes was 150 liters per capita per day. Now it is about 250 ℓpcpd (liters per capita per day). By the year 1985 the water consumption may be 300 ℓpcpd, and by the year 2000 it may reach 400 ℓpcpd.[2] The industrial water requirement is increasing at an average rate of 3 percent per year. A similar situation exists in the U.S.A. In 1965 the per capita consumption in the U.S.A. for domestic, public, and commercial purposes was 460 liters per day (ℓpd), and the industrial requirement was 137 ℓpd; by 1985 it may increase to 465 and 152 ℓpd respectively. Estimations are that the respective increase by the year 2000 will be 474 ℓpcpd and 163 ℓpcpd.[3] The overall rate of increase may be taken as 0.3 percent per year.

In developing countries the rate of increase of water consumption may be much higher than for highly industrialized society. With this prospect, we must regard water as a resource and take proper care to prevent the possible

1

development in the future of serious problems similar to those observed now in the western countries.

Efforts are being made to use sea water which is in abundance. At present the cost of the conversion of salt water prevents any widespread use. This source of useful water is still in the developing stage. Water available below the ground or on the surface must be tapped. The basic requirement for water is that it be adequate and safe for consumption. Water, if contaminated, can cause diseases. Many pathogens are water-borne, and therefore water should be supplied only after proper treatment. In the past, water was either drawn from wells or brought from rivers in buckets or pitchers. There was little idea of a water distribution system for a community. Now units for water supply in individual houses are giving way to a central water supply system which can ensure adequate and safe water. The crux of the problem lies in the economics of collection, treatment, and distribution of water to the community. This book deals with these problems which are of interest to civil and sanitary engineers.

2. Demand for Water

2.1 WATER REQUIREMENT

The water requirement for a particular community depends upon the following factors:

1.   Population
2.   Climatic condition
3.   Habit and mode of living
4.   Plumbing facilities
5.   Sewerage system
6.   Industry
7.   Water tax

In places where meters are installed, the consumption of water is less than when compared to places where a flat rate is charged.

The total water consumption may be estimated by adding domestic consumption, water required for public use, and the water needed by industry. The wastage of water due to leaks, etc., should also be considered when estimating the total demand for water. This may be taken as 10 percent of the total water demand.

On the basis of water consumption per capita per day, the average domestic water requirement of a community is calculated. The average water consumption may be taken as 135 ℓpcpd for India.[1] Water demand may be as high as 450 ℓpcpd in cities and towns with good plumbing facilities. Future population and growth of the community are also to be taken into account for estimating the water required. Water consumption for various purposes is given in Table 2.1, to serve as a guide in estimating the total water requirement of a community.[2]

For comparison, the average domestic requirements of highly developed countries are given in Table 2.2.[3]

2.2 VARIATION IN WATER CONSUMPTION

Water consumption does not remain constant. Yearly, monthly, weekly, daily, and hourly variations in water consumption are observed. Certain dry years cause more consumption. In hot months more water is consumed in drinking, bathing, and watering lawns and gardens. On holidays and weekends the water consumption may be high. Even during a day water

Table 2.1. Water Requirements[2]

Domestic (requirements in liters)	
Drinking, cooking, cleaning per capita per day	20 to 30
One shower bath	40 to 80
One flushing of toilet	8 to 15
Gardening per m²	1 to 2
Public (requirements in liters)	
Street washing per m²	1
Watering plants and green areas per m²	2
School for each child per day	2
Hospitals for each bed per day	200 to 650
Market place per m² per day	5
Office for each worker per day	2
Water treatment plant for backwashing the filter and cleaning the pipes, etc., per capita per day	1 to 2
Commercial (requirements in liters)	
Hotels per bed occupied per day	100 to 150
Slaughterhouses per animal per day	600
Laundry per kg of dry clothes	40 to 70
Industrial (requirements in m³)	
Coal per ton	1.5 to 2.5
Petroleum refinery per 1000 liters	7 to 70
Steel per ton	8 to 20
Artificial fabrics per ton	400
Paper per ton	400 to 600
Sugar per ton of beet	1 to 17
Dairy per 1000 liters of milk	2 to 6
Tannery per skin	1 to 2
Automobile per unit	150
Textile per ton	230 to 270
Gas plants per 1000 m³ gas	7

Table 2.2. Water Requirement of Highly Developed Countries[3]

Country	Population (million)	Average Domestic Requirement (liters per capita per day)
West Germany	59.9	99
Holland	12.7	109
France	49.5	133
Switzerland	6	272
U.S.A.	200	250

use varies with high use during morning hours and close to noon and low use at night.

Generally in smaller communities the variation is large. In bigger cities and towns the demand for water tends to be close to the average. The following values based on experience may be used for estimating the maximum water consumption.

Peak value/day   =   1.2 to 2 times average/day
Peak value/hour =   2 to 3 times average/hour

2.3 FIRE DEMAND

A certain quantity of water must be provided for fighting fire. The total quantity of water required for fire fighting may be a small fraction of the total consumption in large cities, but its effect may be felt in the smaller communities. During a fire the rate of demand of water is high, an important consideration when selecting the size of pipes, pumps, and supply pressures. The basis for fixing the minimum limit of fire demand is the rate of supply and the quantity of water needed to fight the largest possible fire in a city. The quantity of water required to extinguish a fire may be between 200 and 300 m^3, based on average rates of supply of 9 ℓ/s to 12 ℓ/s for a period of 5 to 10 hours.[*] Hydrants should be provided at suitable places in the locality (sect. 7.7). The supply pressure should be about 4 atmosphere when boosters are provided with the fighting squad. Should the rate of supply be guaranteed in the whole of the locality, the minimum diameter of the pipe in the network carrying water should not be less than 100 mm (4 in.). Larger diameter pipes are recommended in the U.S.A.

2.4 POPULATION FORECASTING

2.4.1 Basis of Population Forecast

Any water supply system must be planned to serve the present as well as the future needs of the community. Therefore future population must be assessed while designing the water supply. This requires foresight and judgment. The population changes due to 1) births, 2) deaths, and 3) migration.

Birth, death, and migration rates are dependent on many factors. It is for the forecaster to judge which factors are to be considered in predicting the future population.

Some of the methods of forecasting population are discussed as follows:

---

[*]For central, congested, high value districts of North America, the following formula is applicable for fire demand in a community of 200,000 people or less.

$$Q = 3860 \sqrt{P} \ (1 - 0.01 \sqrt{P})$$

in which   Q = fire demand in l/min
              P = population in thousands

2.4.2 Methods of Forecasting Population[4]

2.4.2.1 Graphical Method

The population during the past years is plotted against time. The forecaster, using his judgment, extends the graph into the future in a manner which fits the trend of population growth in the past. By means of such a graphical extrapolation, future population may be predicted.

2.4.2.2 Arithmetical Method

In this method it is assumed that the rate of population change has been and will remain constant. Expressing it mathematically we have,

$$\frac{dP}{dt} = K_a \quad \cdots \cdots \cdots \cdots \cdots \cdots \quad (2.1)$$

$$dP = K_a dt$$

Here dP/dt represents the change in population P in unit time and K_a is an arithmetic constant.

Integrating between the initial population P_i at the initial year t_i and population P_f at the future year t_f, we get,

$$\int_{P_i}^{P_f} dP = K_a \int_{t_i}^{t_f} dt$$

$$(P_f - P_i) = K_a (t_f - t_i)$$

or $\quad P_f = P_i + K_a (t_f - t_i) \quad \cdots \cdots \cdots \cdots \quad (2.2)$

$$\therefore \quad K_a = \frac{P_f - P_i}{t_f - t_i} = \frac{P_i - P_e}{t_i - t_e} \quad \cdots \cdots \cdots \cdots \quad (2.3)$$

where P_e is the population in some earlier year t_e. From Equation 2.2 it is clear that the relationship between time and population is a straight line and the slope of the line will give the value of K_a.

2.4.2.3 Geometrical Method

It is assumed that the rate of population change is equivalent to the population at a given instant, i.e.

$$\frac{dP}{dt} = k_g P \quad \cdots \cdots \cdots \cdots \cdots \cdots \quad (2.4)$$

in which
$\quad k_g \quad$ is a geometric constant.

Integrating between the initial population P_i at the year t_i and population P_f at the forecast year t_f, we have,

$$\int_{P_i}^{P_f} \frac{dP}{P} = k_g \int_{t_i}^{t_f} dt$$

$$\ln \frac{P_f}{P_i} = k_g \, (t_f - t_i)$$

$$\ln P_f = \ln P_i + k_g \, (t_f - t_i) \quad . \quad . \quad . \quad . \quad . \quad . \quad . \quad . \quad . \quad . \quad (2.5)$$

$$k_g = \frac{\ln P_f - \ln P_i}{t_f - t_i} = \frac{\ln P_i - \ln P_e}{t_i - t_e} \quad . \quad . \quad . \quad . \quad . \quad . \quad . \quad (2.6)$$

where P_e is the population at some earlier year t_e. Equation 2.5 indicates that if population is plotted on a logarithmic scale and time is plotted on a linear scale, a straight line will be obtained. The slope of the graph will give the value of k_g.

Before choosing either the arithmetical or the geometrical method, the past population values should be plotted against time on ordinary graph paper (Figure 2.1). If the relationship between population and time is approximately linear, then the arithmetical method should be used for forecasting the population. If the graph is concave upwards, then the geometrical method may be employed. The time interval selected for finding the values of k_a and k_g may be either the last census interval, or an average of several intervals, or any other selection deemed desirable.

2.4.2.4 Comparative Method

The future population can be predicted by plotting the population of several cities having a similar pattern of growth. The population of the city under study is expected to grow in a similar manner to other older and larger cities. The forecast is made by extrapolating the population curve of the city under study into the future, according to the trend of the population curve of the other cities (Figure 2.2).

2.4.2.5 Ratio and Correlation Method

The growth of a smaller area is closely related to the growth of the population of the region in which the smaller area is situated. Therefore the future population of the smaller area can be estimated by using the forecast of the future population of the region. The census department usually forecasts the future population of the region. With the help of these values one can estimate the future population of the area under study. One can safely assume that,

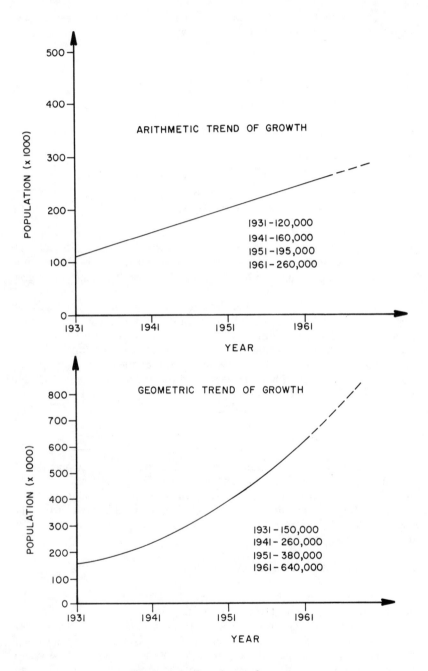

Figure 2.1. Population forecast.

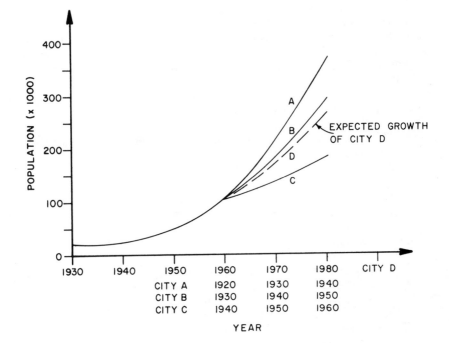

Figure 2.2. Population forecasting by comparative method.

$$\frac{P_f}{P_f{}'} = \frac{P_i}{P_i{}'} = k\ (\text{constant}) \ . \ . \ . \ . \ . \ . \ . \ . \ . \ . \ . \ . \quad (2.7)$$

in which

P_f	=	population forecast for the area under study
$P_f{}'$	=	future population for the region
P_i	=	population of the area under study at the last census
$P_i{}'$	=	population of the region at the last census

If past records for a number of years are available, the ratio k may be taken as an average of these values.

2.4.2.6 Component Method

The following are the main reasons for population change:
1. Birth
2. Death
3. Migration

Where information regarding births and deaths is available, the natural increase can be easily estimated. When calculating, the net migration should be calculated first, otherwise it will not affect the number of births and deaths recorded.

The main advantage in this method is that it is possible to judge separately the influence of various factors on the population growth. This method is particularly useful where migration is not the main factor in population change, because it is often difficult to determine the net migration.

2.4.3 Factors Affecting the Forecast of Population Are:

1. The period of forecast (when the period increases, the accuracy decreases)
2. The population of the area (when the population decreases the accuracy decreases)
3. The rate of increase of population (when the rate increases, the accuracy decreases).

2.4.4 Design Period

The number of years from the date of design to the estimated date when condition of design will be reached is the design period.
For a water supply treatment plant the period is 25 to 50 years.
For a water distribution system the period is 50 years.
Factors affecting the design period are:
1. The usual life of structure and equipment used
2. Initial capital operation and maintenance cost
3. The availability of capital
4. The availability of the expected water consumption at the end of the design period
5. The possibility of extending the plant or increasing its existing capacity
6. The change in the purchasing power of money during the design period

Example 2.1

The census record of a city is given below. Estimate the population of the city in 1970 assuming (a) arithmetic trend in growth (b) geometric trend in growth.

Year	Population (in 1000)
1930	62
1940	74
1950	85
1960	100

a)   For 1930 and 1940 Census records,

$$K_a = \frac{74,000 - 62,000}{10} = 1200$$

For 1940 and 1950 Census records,

$$K_a = \frac{85,000 - 74,000}{10} = 1100$$

For 1950 and 1960 Census records,

$$K_a = \frac{100,000 - 85,000}{10} = 1500$$

$$\therefore \quad \text{Average } K_a = \frac{1200 + 1100 + 1500}{3} = 1266.66$$

We know, $P_f = P_i + K_a (t_f - t_i)$

$$P_{1970} = 100,000 + 1266.66 (1970 - 1960) = 112,666 \text{ or } 112,670$$

b)   For 1930 and 1940 Census records,

$$K_g = \frac{\ln \left(\frac{74}{62}\right)}{10} = 0.01769$$

For 1940 and 1950 Census records,

$$K_g = \frac{\ln \left(\frac{85}{74}\right)}{10} = 0.0138$$

For 1950 and 1960 Census records,

$$K_g = \frac{\ln \left(\frac{100}{85}\right)}{10} = 0.0163$$

Average $K_g = 0.0159$

$\ln P_f = \ln P_i + K_g (P_f - P_i)$

$\ln P_{1970} = \ln 100,000 + 0.0159 \times 10$

$\log P_{1970} = 5.069$

$P_{1970} = 117,200$

Example 2.2

Census records for city A at 10 year intervals are given below. The population of the city in 1960 was 120,000. Calculate the population in 1970 and 1980, assuming (a) an arithmetic trend, and (b) a geometric trend in rate of growth using the least squares method.

Year	Population in 1000 x	Rate of Growth Every Ten Years $y\%$	x^2	xy
1930	60	16.67	3600	1000.2
1940	70	21.43	4900	1500.1
1950	85	11.76	7225	999.6
1960	95	26.32	9025	2500.4
Total	310	76.18	24750	6000.3
Mean	77.5	19.045	6187.5	1500.075

Assuming arithmetic trend of growth.

$$y = a + bx$$

$$a + b\ \Sigma\ \frac{x}{n} - \Sigma\frac{y}{n} = o$$

$$a\Sigma\ \frac{x}{n} + b\Sigma\ \frac{x^2}{n} - \Sigma\ \frac{xy}{n} = o$$

$$a + 77.5\ b - 19.045 = o \quad \ldots \ldots \ldots \ldots \ldots \ldots \ldots \ldots \quad (i)$$

$$77.5\ a + 6187.5\ b - 1500.08 = o \quad \ldots \ldots \ldots \ldots \ldots \quad (ii)$$

Multiplying Equation i by 77.5, we get

$$77.5\ a + 6006.25\ b - 1475.99 = o$$

$$\therefore b = \frac{24.0875}{181.25} = 0.1328$$

$$a + 10.29 - 19.045 = o$$

$$a = 8.755$$

$$y = 8.755 + 0.1328\ x$$

Rate of growth after 1960

$$y = 8.755 + 0.1328 \times 120$$

$$= 8.755 + 15.936$$

$$= 24.691\ \%$$

$\therefore$ Increase in population $= 120 \times \dfrac{24.691}{100} = 29.6292$ thousands

Population in 1970 $= 120 + 29.63 = \underline{149.63}$ thousands.

Rate of growth after 1970

$$y = 8.755 + 0.1328 \times 149.63$$

$$= 8.755 + 19.871 = 29.626\ \%$$

$\therefore$ Increase in population $= 149.63 \times \dfrac{29.626}{100} = 44.32$

Population in 1980 $= 149.63 + 44.32 = \underline{\underline{193.95}}$ thousands.

(b)   Calculation for geometric trend of growth

Population in 1000 x	Rate of Growth Every Ten Years y %	x^2	log y	x log y
60	16.67	3600	1.22194	73.3164
70	21.43	4900	1.33102	93.1714
85	11.76	7225	1.07041	90.9848
95	26.32	9025	1.41995	134.8953
Total 310	76.18	24750	5.04332	392.3679
Mean 77.5	19.045	6187.5	1.26083	98.0919

$$A + 77.5 \ B - 1.26083 = o \quad \dots \dots \dots \dots \dots \dots \quad \text{(iii)}$$

$$77.5 \ A + 6187.5 \ B - 98.0919 = o \quad \dots \dots \dots \dots \dots \quad \text{(iv)}$$

Multiplying Equation iii by 77.5, we get

$$77.5 \ A + 6006.25 \ B - 97.7143 = o$$

$$77.5 \ A + 6187.5 \ B - 98.0919 = o$$

$$- 181.25 \ B = - 0.3776$$

$$B = \frac{0.3776}{181.25} = 0.00208$$

$$A = 1.26083 - 77.5 \ x \ 0.00208$$

$$= 1.26083 - 0.1612 = 1.09963$$

$$\therefore \ \log y = A + Bx = 1.09963 + 0.00208x$$

Rate of population growth after 1960

$$\log y = 1.09963 + 0.00208 \ x \ 120$$

$$= 1.09963 + 0.2496$$

$$= 1.34923$$

$$y = 22.348\%$$

$\therefore$ Population increase in 10 years

$$120 + 120 \ x \ \frac{22.348}{100} = 120 + 26.82 = 146.82 \ \text{thousands}$$

Rate of population growth after 1970

$$\log y = 1.09963 + 0.00208 \ x \ 146.82$$

$$= 1.09963 + 0.30538$$

$$= 1.40501$$

$$y = 25.4103\%$$

Population increase during next 10 years.

$$146.82 + 146.82 \ x \ \frac{25.41}{100} = 146.82 + 37.31 = 184.13 \ \text{thousands}.$$

3. Sources of Water

3.1 WATER CYCLE

Water is available in the following states:
  a. Solid (ice)
  b. Liquid (water)
  c. Gaseous (vapor)

Variations in the amount of energy absorbed from the sun changes the state of water resulting in the water cycle (Figure 3.1). Due to evaporation, a huge quantity of water enters the atmosphere from the earth's surface, from the leaves of trees, and from the surface of water reservoirs. The evaporated water returns to the earth in the form of rain, hail, and snow. A portion of the rain water again evaporates, some infiltrates into the soil, and the rest flows over the surface.

For a particular area the above mentioned process can be represented by the following relationship:

$$P + I_s + I_g = O_s + O_g + E \pm X \; \text{(mm/year)} \quad \ldots \ldots \ldots \quad (3.1)$$

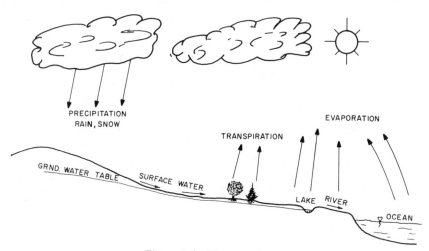

Figure 3.1. Water cycle.

15

in which

P = depth of precipitation

I_s, I_g = depth of surface inflow and infiltration, respectively

$O_s, \ddot{O}_g$ = depth of surface outflow and subsurface seepage, respectively

E = depth of evaporation

X = depth of water in the form of ice, snow, and storage reservoir

A portion of the groundwater or the surface water is utilized by humans for domestic and other purposes which cause changes in the natural water cycle.

3.2 GROUNDWATER

There are two sources of groundwater:

1. Rainfall which permeates into the ground through the pores or cracks in the rock formation and finally reaches the underground water table

2. The water from streams, lakes, and reservoirs which permeates through the soil to the underground water table. Stored or pellicular water, vadose water, and capillary water are not included as groundwater

The capillary fringe which contains capillary water varies in thickness with the change in the water table. Water table in its true sense does not exist in a water bearing stratum found below the impervious stratum.

The intake area of groundwater supplies may be situated very close to the intake point or at a great distance depending upon whether the flow is unconfined or confined within an aquifer (water bearing stratum) below an impervious stratum (Figure 3.2). When the level of the groundwater is free to

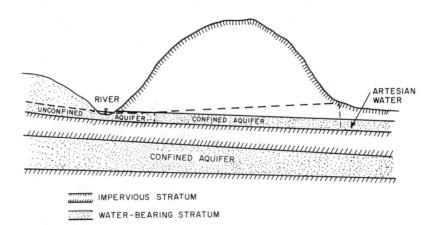

IMPERVIOUS STRATUM

WATER-BEARING STRATUM

Figure 3.2. Occurrence of groundwater.

rise and fall, flow is unconfined or free. In this case the groundwater table follows the slope of the ground surface. The movement of water will then be at right angles to the water table contours. If the aquifer underlies an impervious stratum, flow is confined, which is similar to flow of water in a pipe under pressure.

When the pressure head of the confined water is greater than the distance of water from the ground surface, artesian water will be available. Water found on a lens of impervious stratum above the true water table is called perched water.

A special type of underground water is fresh water found on islands or in the coastal regions near the sea. The fresh water permeates in the porous layer and floats over sea water as an underground lens (Figure 3.3). The fluctuation of fresh water is downward and upward within this lens.

By knowing the height of the water table above the sea level the thickness of the fresh water lens can be determined. For equilibrium the thickness of the fresh water must balance the pressure due to salt water at the bottom of the fresh water.

Therefore we get,

$$HDg = (H - h) D_s g \quad \cdots \cdots \cdots \cdots \cdots \cdots \quad (3.2)$$

in which
H	=	thickness of the fresh water lens (m)
h	=	height of the groundwater table above sea level (m)
D, D_s	=	density of fresh water and salt water respectively (kg/m^3)
g	=	acceleration due to gravity (m/sec^2)

$$(H - H_s) D + H_s D = H_s D_s,$$

in which
H_s	=	height of salt water above the bottom of fresh water lens

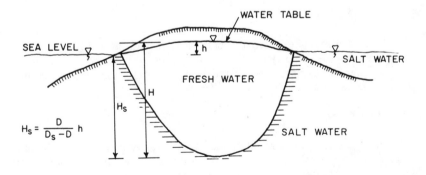

Figure 3.3. Fresh water in contact with salt water adjacent to a coast line.

$$H_s = \frac{D}{D_s - D} (H - H_s)$$

$$= \frac{D}{D_s - D} h \quad \ldots \ldots \ldots \ldots \ldots .(3.3)$$

Assuming $D = 1000$, $D_s = 1025$ ($D_s = 1001.4$ to 1027.7)

$$H_s = 40 \text{ m} \quad \cdots \cdots \cdots \cdots \cdots \quad (3.4)$$

This shows that if the height of the water table is one meter above sea level, the theoretical depth of the fresh water will be 40 meters. Similarly, if pumping is done, the thickness of the fresh water lens will be reduced 40 times the drawdown.

The flow of the groundwater depends upon the hydraulic gradient, type of soil, and soil porosity. Usually the velocity of flow in sand may vary between a fraction of a meter to a few meters per day. The velocity in soil containing gravel may be between 4 to 10 m/day.

While the water flows through the soil it takes with it some of the soluble gases and salts which may be present. This may either increase or decrease the water's usefulness. In general underground water which flows through coal deposits may not be used directly. But water percolating through limestone may be useful due to its chemical content. The filtering capacity of limestone is poor due to cracks and fissures and therefore the quality of water flowing through it may be poor from the bacteriological point of view. Water flowing through gneiss basalt, crystalline slate, or red sandstone does not contain high concentrations of minerals. It is soft and contains carbon dioxide gas. Water found in alluvial sand is polluted because of its proximity to the earth's surface, and it has to be treated before use. Detailed information regarding the geological condition of the area is essential to determine the safety of a water supply from an underground source.

The effect of the water bearing strata on the groundwater quality is given in Table 3.1.

3.3 SURFACE WATER

Surface water is exposed to contamination and invariably it must be treated before use. It may contain both organic and inorganic impurities, gases, and microorganisms. It is generally used for drinking purposes when the groundwater supply is inadequate in quantity.

3.3.1 Rain Water

For municipal water supplies rain water is rarely used directly. It may be used for very small communities which do not have any other source of water supply (deserts) or for areas where water is very hard and unpalatable. Rain water is generally collected from the roofs of buildings and stored in

Table 3.1. Influence of Water Bearing Strata on the Quality of Water (ℓ)

Rocks	Quality of Water
Granite	Dissolved substances negligible—more or less chemically pure
Gneiss	Comparatively larger potash content
Slate	Organic matter in small quantities
Diabase, Syenite Diorite	Slight presence of iron (if rocks are weathered)
Pure quartz	Very shoft water
Pure sand stone	Small quantities of dissolved matter, free from humus, often very pure, soft water
Lime stone	Hard, dissolved matter present
Gypsum and anhydrite	Sulfate is present, hard water
Salt dome	Dissolved matter in large quantities, salty
Marble, Dolomite	Very hard water
Loess, Basic crystalline rock	Rich in dissolved matter especially lime, free from humus

cisterns for domestic consumption. It may also be collected in reservoirs, but large areas where rain water could be collected strictly as rain water may not be available (rain water if allowed to flow over ground will loose its purity). Rain water is soft and it is most suitable for laundry purposes. Due to the absence of minerals, rain water is insipid and when in equilibrium with atmospheric gases (carbon dioxide) it is corrosive.

3.3.2 River Water

Water may be drawn from a river continuously throughout the year or it may be stored during times of flood and then supplied to the community. Because of pollution river water has to be properly treated.

3.3.3 Lakes and Ponds

Deep lakes and ponds having sufficiently large surface areas can be utilized for municipal water supply. The water from these may be better in quality than the river water and can be used for domestic purposes without much treatment. Water can be drawn from deeper layers where the temperature is uniformly low and the plankton life is not dense.

3.3.4 Impounded Reservoirs

During floods and heavy rain, lakes and ponds fill up and discharge to the rivers and streams increases. In the soil adjacent to streams and rivers, water is stored and is called bank storage. The bank storage either supplies water to the water course or receives water from the river depending upon the levels of water the river contains.

The excess amount of water received during rains and floods may be stored, so that it can be used during dry periods. For this purpose an impounded reservoir is constructed by building a dam across a stream. Sometimes storage basins are excavated adjacent to the water course, to store the excess water.

Impounded reservoirs are generally multipurpose and may be constructed to serve the needs of water supply, wastewater disposal, irrigation, flood control, water power, inland transport, aquatic life, fishing, and recreation. Reservoirs are also developed for water supply only.

3.4 SELECTION OF SOURCE

In selecting a source of water the following points should be considered:
1.   The purity of raw water
2.   Volume of available water
3.   Permanency of the sources
4.   Elevation of water level with respect to the area to be supplied
5.   Availability of finances

4. Collection of Water

4.1 GENERAL CONSIDERATION FOR GROUNDWATER

An underground water source can be tapped after detailed geological study and meteorological observations. Soil samples at various depths of borings are taken and analyzed. Longitudinal sections and cross sections of the borings are plotted. The levels of water in the borings are noted at the same time. To find the direction of flow of water, three bore holes should be prepared at each apex of an equilateral triangle (hydrological triangle) having sides of 40 to 60 m in length. The water levels in the bore holes are noted and then plotted (Figure 4.1). By interpolation the groundwater contours are drawn. The groundwater will flow along the line of greatest slope which will be at right angles to the groundwater contours.

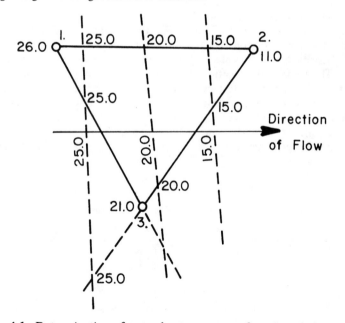

Figure 4.1. Determination of groundwater contours from bore hole water elevations.

21

The yield from an underground aquifer can be determined if the following are known: the thickness of the aquifer, the direction of flow of water, and the velocity of the water. The velocity of the water in the soil strata can be determined by introducing tracers (sodium chloride) or dye, and the measurement can be made at the bore holes. The velocity thus measured will be the velocity through pores. To find the discharge Q, the porosity of soil must also be known.

$$Q = Aev \quad \dots \dots \dots \dots \dots \dots \dots \dots \quad (4.1)$$

in which
- Q = yield of the aquifer (m^3/sec)
- A = cross sectional area of the aquifer (m^2)
- e = porosity (0.25 to 0.5)
- v = velocity inside pores (m/s)

The yield can also be calculated by Darcy's law

$$Q = V_a A = kiA \quad \dots \dots \dots \dots \dots \dots \quad (4.2)$$

in which
- V_a = face or approach velocity (m/s)
- k = coefficient of permeability of the soil (m/s)
- i = hydraulic gradient

The coefficient of permeability k is a constant for a particular soil, and it represents the velocity of flow when the hydraulic gradient is unity. The value of k depends upon the form and size of soil particles, the pore volume and viscosity of water. Values of k for various soils are shown in Figure 4.2.

The value of the coefficient of permeability can be determined either in the field by a pumping test or in the laboratory. Extensive pumping tests in the northern regions of Pakistan and Ghulam Mohammad Barrage area have shown values to lie between 0.3×10^{-2} and 2.2×10^{-2} cm/sec.[1] An average value as high as 9.6×10^{-2} cm/sec has also been reported. Existing wells in an area can also give insight to the nature and specific yield of an underground source of water.

COEFFICIENT OF PERMEABILITY K IN CM/SEC AT UNIT HYDRAULIC GRADIENT

	10^2	10	1.	10^{-1}	10^{-2}	10^{-3}	10^{-4}	10^{-5}	10^{-6}	10^{-7}	10^{-8}	10^{-9}
Nature of Soil		Clean Gravel		Clean Sand; Mixture of Clean Sand and Gravel			Very Fine Sand; Silt; Mixture of Sand, Silt and Clay; Glacial Till; Stratified Clays; etc.				Unweathered Clay	
Flow Characteristics			Good Aquifer					Poor Aquifer			Impervious	

Figure 4.2. Coefficient of permeability for different classes of soils.

4.2 INFILTRATION GALLERIES OR WELLS

Naturally filtered water can be obtained by using galleries. Galleries can be constructed:

a.   Adjacent to a river or artificial recharge basin from which water may be obtained by seepage

b.   Into the sides of hills and mountains

c.   At right angles to artificially built valleys between hills

Infiltration galleries or wells are constructed in the water bearing strata of river beds (Figure 4.3). These can be laid either parallel or at right angles to the river. Construction is of concrete or masonry with perforations. Perforated pipes or pipes with open joints surrounded with layers of graded filter media are employed to collect the water (Figure 4.4).

4.3 DISCHARGE FROM INFILTRATION GALLERIES

Assuming the flow to be uniform and horizontal, the discharge received from one side in an infiltration gallery of length L from an unconfined aquifer can be calculated according to the method given by Dupuit[2] (Figure 4.5).

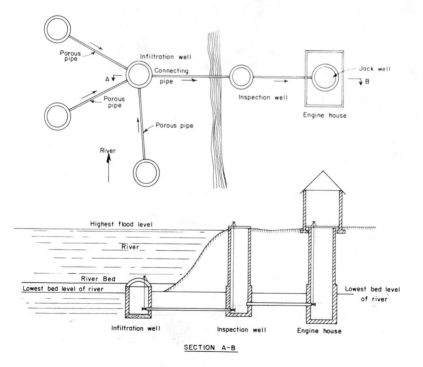

Figure 4.3. Infiltration well.

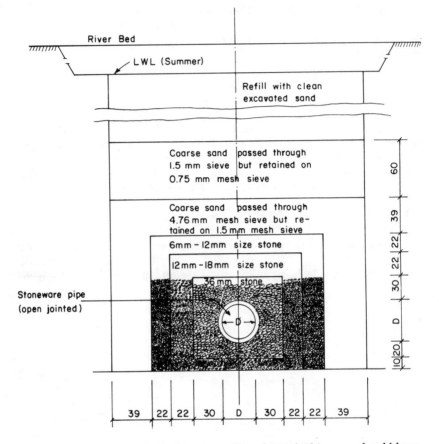

Figure 4.4. Section of an infiltration gallery (pipes laid in rows should have a clearance of 0.61 m, all dimensions are in cms).

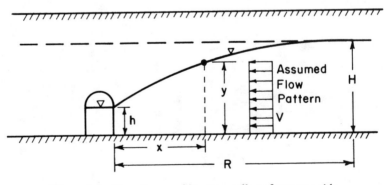

Figure 4.5. Flow in an infiltration gallery from one side.

According to Darcy's law,

$$Q = KiA = K \frac{dy}{dx} L y \quad \ldots \ldots \ldots \ldots \quad (4.3)$$

(See Figure 4.5 for definition of x and y.)
Integrating we get,

$$Q x = K L \frac{y^2}{2} + C$$

for the boundary condition x = o, y = h

$$C = - K L \frac{h^2}{2}$$

$$Q x = K L \frac{y^2}{2} - K L \frac{h^2}{2}$$

when   x = R, y = H

$$Q R = K L \frac{H^2}{2} - K L \frac{h^2}{2}$$

$$= \frac{K L}{2} (H^2 - h^2)$$

$$Q = \frac{K L (H^2 - h^2)}{2 R} \quad \ldots \ldots \ldots \ldots \quad (4.3)$$

If water is entering the gallery from both sides, then L is replaced by 2L in Equation 4.3 and consequently the discharge will be

$$Q = \frac{K L (H^2 - h^2)}{R} \quad \ldots \ldots \ldots \ldots \quad (4.4)$$

The maximum discharge for a perforated pipe of length L laid in flowing underground water of slope S will be

$$Q_{max} = K S L h, \quad \ldots \ldots \ldots \ldots \quad (4.5)$$

where
    h   is the height of water in the pipe.

4.4 UNCONFINED STEADY FLOW IN WELLS

For calculating the discharge from a well driven to the bottom of an aquifer, the principles outlined by Dupuit[2] in conjunction with Darcy's law can be used (Figure 4.6). The underlying assumption is that flow is steady and takes place through openings provided in the well throughout the depth of the aquifer. The water level is assumed to be parallel to the impervious layer at the bottom. When water is withdrawn from the well, it flows radially from concentric vertical cylinders in which the velocity vectors are constant and parallel. After steady state has been reached, the water level depresses towards the well in the form of a curve, called the cone of depression or drawdown curve (Figure 4.6).

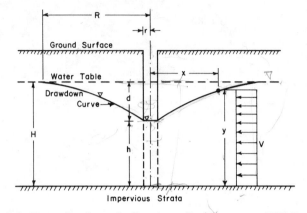

Figure 4.6. Unconfined steady flow in well. (Courtesy of TH-Dresden.)

According to Darcy's law (Equation 4.2)

$$Q = KiA$$

$$i = \frac{dy}{dx} = \text{slope of the drawdown curve at a point}$$

(In Darcy's law, i is generally considered to be equal to the sin θ, but here the tangent of the horizontal angle has been considered which is permissible for small angles.)

$A = 2\pi\,x\,y$ = Area of the outer circular boundary which is concentric with the well and the area through which the flow passes.

$$Q = K \frac{dy}{dx}\, 2\pi\,x\,y$$

$$2y\,dy = \frac{Q}{\pi K}\, \frac{dx}{x}$$

Integrating we get

$$y^2 = \frac{Q}{\pi K}\ln x + C$$

when $\quad\quad y = h$, and $x = r$ (radius of the well)

$$C = h^2 - \frac{Q}{\pi K}\ln r$$

Hence, $\quad\quad y^2 = h^2 + \frac{Q}{\pi K}\ln r \quad$ . . . . (4.6)

which is the equation of the drawdown curve.

Also $\quad\quad Q = \dfrac{\pi K\,(y^2 - h^2)}{\ln \dfrac{x}{r}} \quad$ . . . . (4.7)

when $\quad\quad y = H$, $x = R$ (radius of circle of influence)

The yield of the well is defined as follows:

$$Q = \frac{\pi K \, (H^2 - h^2)}{\ln \frac{R}{r}} \qquad \dots \dots \dots \dots \dots \quad (4.8)$$

The drawdown d at the pumping well is H - h. Solving for h and substituting the value into Equation 4.7 yields

$$Q = \frac{\pi K \, [H^2 - (H - d)^2]}{\ln R/r} \qquad \dots \dots \dots \dots \quad (4.9)$$

Also
$$d = H - \sqrt{H^2 - \frac{Q \ln \frac{R}{r}}{\pi K}} \qquad \dots \dots \dots \quad (4.10)$$

According to Sichardt[3] the radius R of the circle of influence can be estimated by the following expression

$$R \approx 3000 \, d \, \sqrt{K} \qquad \dots \dots \dots \dots \dots \quad (4.11)$$

in which
  d   is the drawdown (m)
  K   is the coefficient of permeability (m/s)
  R   is in m
  R can also be calculated with Equation 4.8

$$\ln R = \frac{\pi K \, (H^2 - h^2)}{Q} + \ln r \qquad \dots \dots \dots \dots \quad (4.12)$$

When the values of Q and K are not known, the value of R may be determined by considering the coordinates of points (r, h; x, y; R, H) on the drawdown curve. From Equations 4.7 and 4.8, we have

$$Q = \frac{\pi K \, (y^2 - h^2)}{\ln \frac{x}{r}} = \frac{\pi K \, (H^2 - h^2)}{\ln \frac{R}{r}}$$

Hence
$$\ln R = \frac{(H^2 - h^2) \ln \frac{x}{r}}{y^2 - h^2} + \ln r \qquad \dots \dots \dots \dots \quad (4.13)$$

4.6 CONFINED STEADY FLOW

The yield from a well driven through a confined aquifer (water bearing strata lying between impervious layers) can be calculated as follows (Figure 4.7)

$$Q = 2\pi \, x \, m \, K \, \frac{dy}{dx}$$

in which
  m   =   thickness of the water bearing strata.

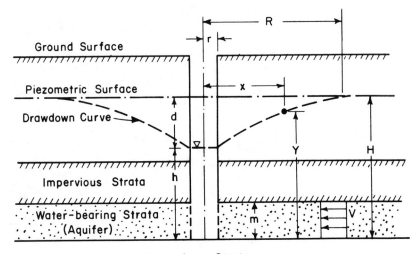

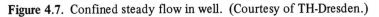

Figure 4.7. Confined steady flow in well. (Courtesy of TH-Dresden.)

Therefore,

$$dy = \frac{Q}{2\pi Km} \frac{dx}{x}$$

Integrating we get,

$$y = \frac{Q}{2\pi Km} \ln x + C$$

When $y = h,\ x = r,$

$$C = h - \frac{Q}{2\pi Km} \ln r$$

Hence $y = h + \dfrac{Q \ln \frac{x}{r}}{2\pi Km}$ . . . . (4.14)

which is the equation of the drawdown curve. Therefore, the yield from the well is defined as follows:

$$Q = \frac{2\pi Km\,(y - h)}{\ln \frac{x}{r}}$$ . . . . (4.15)

when, $y = H,\ x = R:$

$$Q = \frac{2\pi Km\,(H - h)}{\ln \frac{R}{r}}$$ . . . . (4.16)

and $d = H - h = \dfrac{Q \ln \frac{R}{r}}{2\pi Km}$ . . . . (4.17)

4.7 STAGNATION POINT

The equations derived for the confined and unconfined cases are applicable to aquifers with level piezometric surfaces and sloped surfaces, if the undisturbed piezometric surface runs parallel to the bottom impervious layer. In a sloping piezometric surface the flow net does not consist of radial flow lines and concentric equipotential lines as the cone of depression does not remain symmetrical. In Figure 4.8 the cross section at AA which has been taken at right angles to the direction of flow is similar to Figure 4.7 which is for a level piezometric surface; whereas, in section BB the drawdown curve at the upstream lies higher than the curve on the downstream. The flow net for the groundwater having piezometric slope can be drawn by taking the resultant of the linear flow and the radial flow. The apex L_i of the limiting flow path gives the point of stagnation which demarcates the boundary of influence of the well. If this point is projected on the drawdown curve of

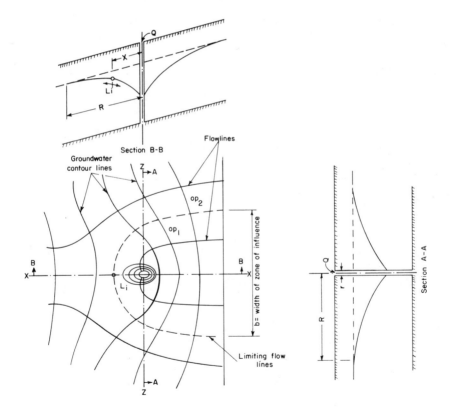

Figure 4.8. Flow net of well in flowing groundwater. (Courtesy of TH-Dresden.)

section BB, the tangent to the drawdown curve at this point will be horizontal. Inside the limiting flow line, which passes through the point L_i and is at right angles to the direction of flow at that point, the water particles (e.g. P_i) flow towards the well, while particles (e.g. P_2) outside the limiting flow line flow past the well.

We know that

$$Q = \frac{2\pi Km\ (H - h)}{\ln \frac{R}{r}} \qquad \ldots \ldots \ldots \ldots \quad (4.16)$$

Flow through unit depth of aquifer is

$$q = \frac{Q}{m} = \frac{2\pi K\ (H - h)}{\ln \frac{R}{r}} \qquad \ldots \ldots \ldots \ldots \quad (4.17)$$

The velocity potential for radial flow is

$$\phi_r = v\,x = K\,i\,x = K\frac{H - h}{x}\,x = K\,(H - h)$$

From Equation 4.18 we have,

$$K\,(H - h) = \frac{q}{2\pi} \ln \frac{R}{r} \qquad \ldots \ldots \ldots \ldots \quad (4.18)$$

In this case $R^2 = X^2 + Z^2$

$$\therefore \phi_r = \frac{q}{2\pi} \ln \frac{R}{r} = \frac{q}{2\pi} \left[\frac{1}{2}\ln\ (X^2 + Z^2) - \ln r\right]. \quad \ldots \quad (4.19)$$

The velocity potential of the combined flow is the sum of the radial flow and the undisturbed groundwater flow.

$$\phi = \phi_u + \phi_r$$

$$\phi = v_u\,X + \phi_r$$

But

$$v_u = \frac{\partial \phi}{\partial x} = v_u + \frac{q}{2\pi} \left(\frac{X}{X^2 + Z^2}\right) \qquad \ldots \ldots \ldots \quad (4.20)$$

At the point of stagnation downstream from the well the velocity along the x-direction will be zero.

$$v_u + \frac{q}{2\pi} \left(\frac{X}{X^2 + Z^2}\right) = 0$$

When $Z = 0$,

$$v_u = -\frac{q}{2\pi} \times \frac{1}{X}$$

or

$$X = -\frac{q}{2\pi v_u} \qquad \ldots \ldots \ldots \ldots \ldots \quad (4.21)$$

But

$$\frac{q}{v_u} = b = \text{width of zone of influence}$$

$$\therefore b = -2\pi x \qquad \ldots \ldots \ldots \ldots \ldots \quad (4.22)$$

By knowing b, the width of the zone of influence, one can establish the proper spacing between two pumping wells. It is advisable to have the distance between the wells always greater than the width of the zone of influence.

4.8 SPECIFIC CAPACITY AND DEVELOPMENT OF WELLS

The specific capacity of a well, which is the yield per unit drawdown, can be determined by pumping for an extended period of time. Before making any observation the wells should be properly flushed and cleaned. All these operations are called well development. The discharge from the well is measured when the pumped water is found free of sand particles. Pumping should continue long enough to reach a stable water level. When the drawdown reaches a fairly stable value, further increase in the yield is unlikely. If the rate of pumping is increased, the yield may increase for a short while after which the pump filter may become clogged due to the high velocity of the water impinging material on the filter.

By examining the equations in sections 4.5 and 4.6, it is evident that the increase in the diameter of the well will not cause a considerable increase in the yield. Increasing the diameter of a well from 8 cm to 60 cm, may produce an increase in yield of 15 to 30 percent only.

4.9 FEASIBILITY OF GROUNDWATER DEVELOPMENT

The groundwater may be considerably lowered due to indiscriminate construction of wells and pumping of groundwater. Pumping costs increase as the water table is lowered. The following points should be considered in groundwater development decisions:

1. Water must be available in sufficient quantity
2. Depth of water should be such that the pumping is economical
3. The water table should be stable
4. The withdrawals may not be in excess of the natural recharge

4.10 PUMPING TEST FOR SOIL PERMEABILITY

A pumping test is used to determine the permeability of soil. Observation wells are sunk at distances varying between 5 and 50 m from the test well to measure the drawdown (Figure 4.9).

Applying Darcy's law for unconfined flow we get,

$$\int_{y_1}^{y_2} y \, dy = \frac{Q}{2\pi K} \int_{x_1}^{x_2} \frac{dx}{x}$$

$$K = \frac{Q \ln \frac{x_2}{x_1}}{\pi (y_2{}^2 - y_1{}^2)} \quad \ldots \ldots \ldots \ldots \ldots \ldots (4.23)$$

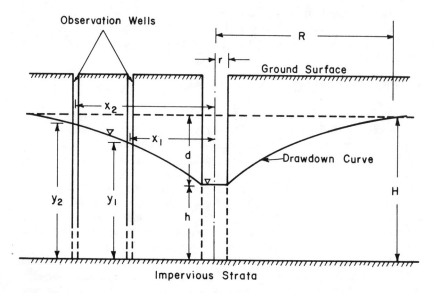

Figure 4.9. Arrangement for pumping test. (Courtesy of TH-Dresden.)

Similarly for confined flow we have,

$$K = \frac{Q \ln \frac{x_2}{x_1}}{2\pi m (y_2 - y_1)} \quad \dots \dots \dots \dots \dots \dots \quad (4.24)$$

Example 4-1

A well is driven in an aquifer 10 m thick and draws water laterally from a distance of 17 m. The piezometric surface is lowered. The gradient of the undisturbed piezometric surface is 1:3.5. Calculate the rate of flow through the aquifer. Assume K (coefficient of permeability) = 0.36 m/hr.

$$Q = KiA = 0.36 \frac{1}{3.5} \times (2 \times 10 \times 17)$$

$$= 0.36 \times \frac{1}{3.5} \times 340 = 35 \text{ m}^3/\text{hr}.$$

Example 4-2

The diameter of a well is 0.50 m in which there is unconfined steady flow. The well penetrates fully into the impervious layer, and has a drawdown of 5 m. The height of the free water surface is 50 m above the impervious layer. Assuming the coefficient of permeability is 0.85 m/hr, calculate the yield from the well.

COLLECTION OF WATER 33

$$Q = \frac{\pi K (H^2 - h^2)}{\ln \frac{R}{r}} \quad \cdots \cdots \cdots \cdots \cdots \cdots \quad (4.8)$$

Here
H = 50 m
h = 45 m
r = 0.25 m
K = 0.85 m/hr = 2.36 x 10^{-4} m/sec.

From Equation 4.11

$$R = 3000\, d \sqrt{K} = 3000 \times 5 \sqrt{2.36 \times 10^{-4}}$$

$$= 3000 \times 5 \times 1.535 \times 10^{-2} = 230\ m$$

$$\therefore Q = \frac{\pi \times 0.85\,(50^2 - 45^2)}{\ln \frac{230}{0.25}} = \frac{2.67 \times 475}{2.303 \log 920}$$

$$= \frac{2.67 \times 475}{2.303 \times 2.9638} = 185\ m^3/hr.$$

Example 4-3

A well of 0.6 m diameter is driven in an aquifer having unconfined flow (Figure 4.10). The water level when not being pumped is 30 m above the impervious layer. When water is pumped at 150 m^3/hr the level of water below the original level in the observation well at a distance of 15 m is 3 m and in another well at a distance of 30 m is 1 m. Calculate the drawdown at the pumping well.

Q = 150 m^3/hr
x_2 = 30 m; $\ln 30 = \ln 3.0 + \ln 10 = 1.0986 + 2.3026 = 3.4012$
x_1 = 15 m; $\ln 15 = \ln 1.5 + \ln 10 = 0.4055 + 2.3026 = 2.7081$
y_2 = 30-1 = 29 m, y_1 = 30-3 = 27 m

From Equation 4.23

$$K = \frac{Q \ln \frac{x_2}{x_1}}{\pi (y_2^2 - y_1^2)} = \frac{150\,(3.4012 - 2.7081)}{3.14\,(29 + 27)(29 - 27)} = \frac{150\,(0.6931)}{3.14\,(56)(2)}$$

$$= 0.295\ m/hr$$

Also

$$Q = \frac{\pi K (H^2 - h^2)}{\ln \frac{R}{r}} = \frac{\pi K (y_2^2 - y_1^2)}{\ln \frac{x_2}{x_1}}$$

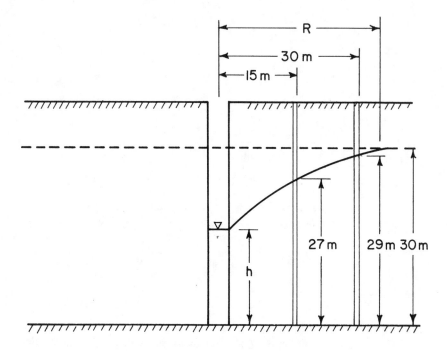

Figure 4.10. Example 4.3.

The radius of the zone of influence can be calculated by considering the distance between the point where the drawdown curve begins and one of the test wells. In this example, the test well near the pumped well is used. The radius of the zone of influence can be substituted for x_2 in the above equation because at the end of the zone of influence $x_2 = R$.

$$\therefore 150 = \frac{0.295 \, \pi \, (30^2 - 27^2)}{\ln \dfrac{R}{15}}$$

Solving we get

$$R = 43.11 \text{ m}$$

$$\therefore H^2 - h^2 = \frac{Q \ln \dfrac{43.11}{0.3}}{\pi \, (0.295)} = 804$$

$$h^2 = (30)^2 - 804 = 96$$

$$h = 9.8 \text{ m}$$

$$\therefore \text{ drawdown} = 30 - 9.8 = 20.2 \text{ m}$$

Example 4-4

Water is pumped from a well at a steady rate of 90 m³/hr. The drawdowns in two observation wells located at 10 m and 25 m from the pumping well are 0.58 and 0.46 m, respectively. The height of water in the layer is 12.00 m. Calculate the permeability of the soil.

$$Q = 90 \text{ m}^3/\text{hr}$$
$$x_2 = 25 \text{ m}, \ln x_2 = 3.21888; \quad y_2 = 12.00 - 0.46 = 11.54 \text{ m}$$
$$x_1 = 10 \text{ m}, \ln x_1 = 2.30260; \quad y_1 = 12.00 - 0.58 = 11.42 \text{ m}$$
$$\ln x_2/x_1 = 0.91628$$

From Equation 4.23

$$K = \frac{Q \ln \frac{x_2}{x_1}}{\pi (y_2{}^2 - y_1{}^2)} = \frac{90 \times 0.91628}{3.14 \times (11.54 + 11.42)(11.54 - 11.42)}$$

$$= \frac{82.4652}{3.14 \times 22.96 \times 0.12} = 9.55 \text{ m/hr}$$

Example 4-5

A well penetrates an aquifer 15 m thick (Figure 4.11) and is pumped, resulting in a lowering of the piezometric surface. The slope of the undisturbed piezometric surface is 1 m in 4 m and the yield of the well is 200 m³/hr. Calculate the distance of the stagnation point from the well and the width of the zone of influence.

Assume $K = 0.9$ m/hr

Yield of the well = flow through the aquifer
Flow through unit depth of aquifer

$$q = \frac{Q}{m} = \frac{Q}{15}$$

The distance of point of stagnation (Equation 4.21)

$$x = \frac{q}{2\pi v_u} = \frac{Q}{15 \times 2\pi v_u}$$

But

$$v_u = Ki = 0.9 \times \frac{1}{4}$$

$$\therefore x = \frac{200 \times 4}{30 \times 0.9} = 9.45 \text{ m}$$

Width of zone of influence

$$b = 2\pi x = 2 \times 3.14 \times 9.45 = 59.2 \text{ m}$$

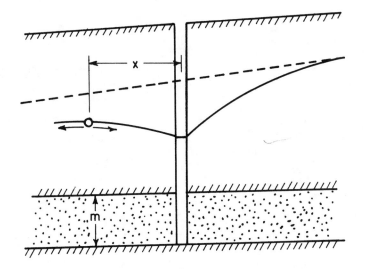

Figure 4.11. Example 4.5.

4.11 GROUP OF WELLS

Let n be the number of wells and Q be the total yield of the wells. Therefore, the mean yield from each well is Q/n. The principle of superposition is used to get the following relationship based on Equation 4.6 for an unconfined aquifer.

$$y^2 - h^2 = \frac{Q}{2\pi K} \left(\ln \frac{x_1}{r_1} + \ln \frac{x_2}{r_2} + ... + \ln \frac{x_n}{r_n} \right) \quad . . . . . (4.25)$$

in which $x_1, x_2, x_3 ... x_n$ is the distances of the wells 1, 2, 3 ... n from an observation well having y as the height of water level and h as the height of water in an arbitrary pumping well of radius r and discharging at a rate of Q. Assuming that the observation well is far from the wells, so that $x_1 = x_2 = x_3$ ... x_n is approximately correct where x can be assumed to be the distance between the center of gravity of the group of wells. Further we know that the logarithms of large numbers do not differ significantly, hence Equation 4.25 can be written as follows:

$$y^2 - h^2 = \frac{Q}{n\pi K} n \ln \frac{x}{r} = \frac{Q}{\pi K} \ln \frac{x}{r} \quad . . . . (4.26)$$

in which

$$n \ln r = \ln r_1 + \ln r_2$$

$$\ln r^n = \ln [(r_1)(r_2) \dots]$$

$$\therefore r^n = r_1 \cdot r_2 \dots \qquad \dots \qquad (4.27)$$

Knowing the value of h, the height of water levels in each well can be computed as follows:

For well No. 1 (Figure 4.12)

$$h_1^2 - h^2 \frac{Q}{n\pi K} \left(\ln \frac{x_{12}}{r_2} + \ln \frac{x_{13}}{r_3} + \dots \right)$$

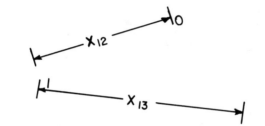

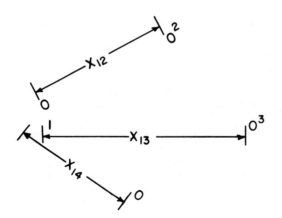

Figure 4.12. Group of wells.

$$= \frac{Q}{2\pi K} \ln \frac{x_{12} \cdot x_{13}, \; \cdots}{r_2 \cdot r_3 \; \cdots} \qquad \cdots \cdots \cdots \cdots (4.28)$$

in which

h_1 is the height of water level in well 1, $x_1 = r_1$, and assuming $x_2 = x_{12}$; $x_3 = x_{13} \; \cdots$ and so on.

Let there by two observation wells a and b at a distance of $a_1, b_1, \ldots$ and so on.

For the well a we get,

$$y_a^2 - h^2 = \frac{Q}{n\pi K} \ln \frac{(x_{a1} \cdot x_{a2} \; \cdots)}{(r_1 \cdot r_2 \; \cdots)} \qquad \cdots \cdots \cdots (4.29)$$

Similarly, for well b we get

$$y_b^2 - h^2 = \frac{Q}{n\pi K} \ln \frac{x_{b1} \cdot x_{b2} \; \cdots}{r_1 \cdot r_2 \; \cdots} \qquad \cdots \cdots \cdots (4.30)$$

Hence from Equations 4.29 and 4.30 we get

$$y_a^2 - y_b^2 = \frac{Q}{2\pi K} \ln \frac{x_{a1} \cdot x_{a2} \; \cdots}{x_{b1} \cdot x_{b2} \cdots} \qquad \cdots \cdots \cdots (4.31)$$

Similarly, an expression for a confined aquifer can be written

$$y_a - h = \frac{Q}{2\pi K mn} \ln \frac{x_{a1} \cdot x_{a2} \; \cdots}{r_1 \cdot r_2 \; \cdots} \qquad \cdots \cdots \cdots (4.32)$$

Hence for two observation wells a and b we get

$$y_a - y_b = \frac{Q}{2\pi K mn} \ln \frac{x_{a1} \cdot x_{a2} \; \cdots}{x_{b1} \cdot x_{b2} \; \cdots} \qquad \cdots \cdots \cdots (4.33)$$

4.12 UNSTEADY FLOW IN WELLS

Theis[4] and Jacob,[5] based on a non-equilibrium condition, developed expressions for the flow of water in wells. The underlying assumption is that the well first utilizes the water stored in the vicinity and then gradually draws water through enlargement of the cone of depression.

The expression from Theis is based on the analogy between flow of heat towards a sink or a point where heat is taken away at a uniform rate, and the laminar flow of water. Taking hydraulic pressure analogous to temperature; pressure gradient to temperature gradient; permeability to thermal conductivity; and specific yield (coefficient of storage) to specific heat we get,

$$d = \frac{Q}{4\pi T} f(u) \; \cdots \cdots \cdots \cdots \cdots \cdots \cdots (4.34)$$

in which

d	=	drawdown at any point on the cone of depression (m)
Q	=	uniform rate of pumping (m^3/hr)
T	=	coefficient of transmissibility (m^3/hr·m)
T	=	(KH for unconfined aquifer and km for confined aquifer of thickness m)

$$f(u) = \int_u^\infty \frac{e^{-u}\ du}{u} \quad . \ . \ . \ . \ . \ . \ . \ . \ . \ . \ . \ . \ . \ . \quad (4.35)$$

in which

u	=	$\dfrac{x^2 S}{4Tt}$
x	=	distance of the observation well from the pumped well (m)
S	=	storage coefficient
		(For an unconfined aquifer it corresponds to specific yield, and for a confined aquifer it is the volume of water released from or taken into storage per unit surface area of the column of water per unit drawdown in the piezometric surface.)
t	=	time the well has been pumped (hrs)

Values of "well function" f(u) from Table 4.1 for various values of u are plotted on logarithmic paper (Figure 4.13). The drawdown observations are plotted on the same logarithmic paper versus x^2/t. The two sets of points are joined by smooth curves. The curve obtained by joining the first set of points is called the "Type" curve. The two curves are similar because d and x^2/t are related in the same manner as u and f(u). Therefore, by superimposing the curve of the observed data over the "Type" curve the coordinates of the matching point, i.e. the values of f(u) and u that correspond to the values of d and x^2/t, can be found. In this manner the exponential equation can be solved and the values of T and S can be determined.

T and S values corresponding to the information regarding time and drawdown at the pumped well and at other observation wells can be calculated. The different values of T and S obtained in this manner can be averaged. Using this average value, drawdown in the well at any time can be determined.

The method discussed in the above paragraph is inconvenient. Jacob[5] has suggested a simpler equation which can be derived as follows:

The well function f(u) (Equation 4.35) can be written in the form of the following convergent series:

$$f(u) = -0.5772 - \ln u + u - \frac{u^2}{2!2} + \frac{u^3}{3!3} - \quad . \ . \ . \ . \ . \quad (4.36)$$

Substituting Equation 4.36 in Equation 4.34 we get,

$$d = \frac{Q}{4\pi T} \left(-0.5772 - \ln u + u - \frac{u^2}{2!2} + \frac{u^3}{3!3} \ ... \right) \quad . \ . (4.37)$$

Table 4.1. Values of the Well Function $f(u)$ for Values of u^6

u	1.0	1.5	2.0	2.5	3.0	3.5	4.0	4.5	5.0	5.5	6.0	6.5	7.0	7.5	8.0	8.5	9.0	9.5
$\times 1$	2.194×10^{-1}	1×10^{-1}	4.89×10^{-2}	2.491×10^{-2}	1.305×10^{-2}	6.97×10^{-3}	3.779×10^{-3}	2.073×10^{-3}	1.148×10^{-3}	6.409×10^{-4}	3.601×10^{-4}	2.034×10^{-4}	1.155×10^{-4}	6.583×10^{-5}	3.767×10^{-5}	2.162×10^{-5}	1.245×10^{-5}	7.185×10^{-6}
$\times 10^{-1}$	1.823	1.465	1.223	1.044	0.9057	0.7942	0.7024	0.6253	0.5598	0.5034	0.4544	0.4115	0.3738	0.3403	0.3106	0.2840	0.2602	0.2387
$\times 10^{-2}$	4.038	3.637	3.355	3.137	2.959	2.810	2.681	2.568	2.468	2.378	2.295	2.220	2.151	2.087	2.027	1.971	1.919	1.870
$\times 10^{-3}$	6.332	5.927	5.639	5.417	5.235	5.081	4.948	4.831	4.726	4.631	4.545	4.465	4.392	4.323	4.259	4.199	4.142	4.089
$\times 10^{-4}$	8.633	8.228	7.940	7.717	7.535	7.381	7.247	7.130	7.024	6.929	6.842	6.762	6.688	6.619	6.555	6.494	6.437	6.383
$\times 10^{-5}$	10.94	10.53	10.24	10.02	9.837	9.683	9.550	9.432	9.326	9.231	9.144	9.064	8.990	8.921	8.856	8.796	8.739	8.685
$\times 10^{-6}$	13.24	12.83	12.55	12.32	12.14	11.99	11.85	11.73	11.63	11.53	11.45	11.37	11.29	11.22	11.16	11.10	11.04	10.99
$\times 10^{-7}$	15.54	15.14	14.85	14.62	14.44	14.29	14.15	14.04	13.93	13.84	13.75	13.67	13.60	13.53	13.46	13.40	13.34	13.29
$\times 10^{-8}$	17.84	17.44	17.15	16.93	16.75	16.59	16.46	16.34	16.23	16.14	16.05	15.97	15.90	15.83	15.76	15.70	15.65	15.59
$\times 10^{-9}$	20.15	19.74	19.45	19.23	19.05	18.89	18.76	18.64	18.54	18.44	18.35	18.27	18.20	18.13	18.07	18.01	17.95	17.89
$\times 10^{-10}$	22.45	22.04	21.76	21.53	21.35	21.20	21.06	20.94	20.84	20.74	20.66	20.58	20.50	20.43	20.37	20.31	20.25	20.20
$\times 10^{-11}$	24.75	24.35	24.06	23.83	23.65	23.50	23.36	23.25	23.14	23.05	22.96	22.88	22.81	22.74	22.67	22.61	22.55	22.50
$\times 10^{-12}$	27.05	26.65	26.36	26.14	25.96	25.80	25.67	25.55	25.44	25.35	25.26	25.18	25.11	25.04	24.97	24.91	24.86	24.80
$\times 10^{-13}$	29.36	28.95	28.66	28.44	28.26	28.10	27.97	27.85	27.75	27.65	27.56	27.48	27.41	27.34	27.28	27.22	27.16	27.11
$\times 10^{-14}$	31.66	31.25	30.97	30.74	30.56	30.41	30.27	30.15	30.05	29.95	29.87	29.79	29.71	29.64	29.58	29.52	29.46	29.41
$\times 10^{-15}$	33.96	33.56	33.27	33.05	32.86	32.71	32.57	32.46	32.35	32.26	32.17	32.09	32.02	31.95	31.88	31.82	31.76	31.71

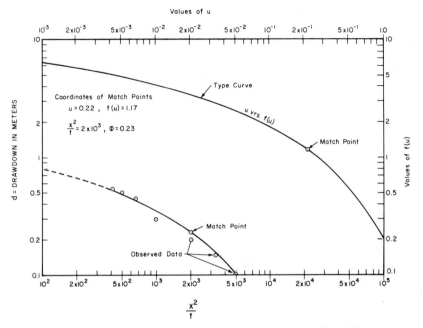

Figure 4.13. Use of non-equilibrium method in a well problem.

Hence
$$d_1 = \frac{Q}{4\pi T}\left(-0.5772 - \ln \frac{x^2 S}{4T}\frac{1}{t_1}\right) \quad \cdots$$

neglecting the remaining terms on the assumption that u is inversely proportioned to t.

Similarly
$$d_2 = \frac{Q}{4\pi T}\left(-0.5772 - \ln \frac{x^2 S}{4T}\frac{1}{t_2}\right)$$

Therefore
$$d_2 - d_1 = \frac{Q}{4\pi T}\ln \frac{t_2}{t_1}$$

Also
$$T = \frac{2.3\,Q}{4\pi(d_2 - d_1)}\log \frac{t_2}{t_1} \quad \cdots\cdots\cdots\cdots (4.38)$$

in which

   T is the coefficient of transmissibility ($m^3/hr\ m$)
   Q is the discharge (m^3/hr)
   t_2 and t_1 are times elapsed in hours since pumping started
   d_2 and d_1 are the corresponding drawdowns in meters in an observation well

Equation 4.38 can be written as

$$T = \frac{2.3Q}{4\pi \Delta d} \log \frac{t_2}{t_1} \quad \ldots \ldots \ldots \ldots \ldots \ldots (4.39)$$

in which

$\Delta d = d_2 - d_1 = $ change in drawdown between t_1 and t_2.

Equation 4.39 can be further simplified by taking the value of Δd for a complete logarithmic cycle, so that

$$\log \frac{t_2}{t_1} = 1 = \log \frac{100}{10} = 1$$

Hence

$$T = 0.183 \frac{Q}{\Delta d} \quad \ldots \ldots \ldots \ldots \ldots \ldots \ldots (4.40)$$

For small values of x and large values of t, the well function can be approximated by the first two terms (Equation 4.37)

$$d = \frac{Q}{4\pi T} \left(-0.5772 - \ln \frac{x^2 S}{4Tt} \right)$$

$$= \frac{2.3\,Q}{4\pi T} \left(\log \frac{2.246Tt}{x^2 S} \right) \quad \ldots \ldots \ldots \ldots \ldots (4.41)$$

when

$$d = o.$$

$$S = \frac{2.246Tt_o}{x^2} \quad \ldots \ldots \ldots \ldots \ldots \ldots \ldots (4.42)$$

in which

t_o is the intercept (in hours) at $d = o$, obtained by extending the straight line portion of the graph between the drawdown and time (Figure 4.14)

The graph between the drawdown and pumping time is not a straight line at the beginning and therefore Jacob's modified Equation 4.40 cannot be applied for periods immediately after the start of pumping. For confined aquifers the curve becomes a straight line in a short time, but for unconfined aquifers it requires a longer period.

The non-equilibrium equations are based on the assumptions that: (1) stream lines are parallel to each other, i.e. there is small drawdown and full penetration of the well in the aquifer, (2) dewatering of the aquifer takes place as soon as the water table drops, (3) the value of T and S are constant.

These assumptions closely approximate confined aquifers, and therefore this method is more suited to confined aquifers. For thin or poorly permeable strata it should be used with care.

Example 4-6

The following observations were made on a well at a distance of 100 m from a pumped well. The rate of pumping is 2 m^3/min.

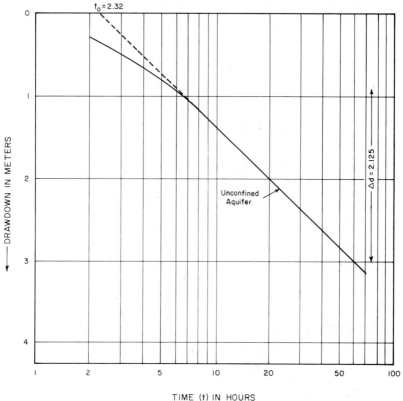

Figure 4.14. Unsteady flow in wells (modified method).

X	Time (hr)	d = Drawdown (m)	x^2/t
100 m	2	0.1	5×10^3
	3	0.15	3.33×10^3
	5	0.2	2×10^3
	10	0.3	1×10^3
	15	0.45	6.66×10^2
	20	0.50	5×10^2
	24	0.55	4.16×10^2

Find the drawdown when pumping at a rate of 4 m³/min for 10 hours.

Values of d and x^2/t are plotted on logarithmic paper, and after that the match point is found by superimposing the "Type" curve, taking care that the coordinate axes remain parallel (Figure 4.13).

The coordinates of the match point are:

$$u = 0.22 = 2.2 \times 10^{-1}; \quad f(u) = 1.17$$

$$\frac{x^2}{t} = 2 \times 10^3, \quad d = 0.23$$

Substituting these values in the equations

$$d = \frac{Q}{4\pi T} f(u)$$

$$0.23 = \frac{(2)(60)}{4(3.14)T} \times 1.17$$

$$T = \frac{120(1.17)}{12.56(0.23)} = 48.6 \text{ m}^3/\text{hr}\cdot\text{m}$$

$$u = \frac{x^2 S}{4Tt}$$

or

$$S = \frac{4uT}{x^2/t} = \frac{4(0.22)(48.6)}{2 \times 10^3} = 21.4 \times 10^{-3}$$

If the pumping is done at the rate of 4 m^3/min for 10 hours then,

$$u = \frac{(100)^2 \times 21.4 \times 10^{-3}}{4 \times 48.6 \times 10} = 1.1 \times 10^{-1}$$

From Table 4.1 the value of $f(u) = 1.75$

$$\therefore d = \frac{Q}{4\pi T} f(u) = \frac{4 \times 60}{4 \times 3.14 \times 48.6} \times 1.75 = 0.69 \text{ m}$$

Example: 4-7

The following information is given regarding an observation well.
1.   Using the modified, or Δd, method calculate the drawdown in the observation well after 60 hours and after 6 months when pumping is done at a uniform rate of 1.25 m^3/min.

Time (hours)	2	4	7	10	20	30	60
Drawdown (m)	0.30	0.745	1.125	1.375	2.0	2.375	3.0

$$Q = 1.25 \text{ m}^3/\text{min} = 75 \text{ m}^3/\text{hr}$$

$$X = 50 \text{ m; from the graph (Figure 4.14) } \Delta d = 2.125 \text{ m},$$
$$t_o = 2.32 \text{ hrs}$$

$$T = \frac{0.183 \, Q}{\Delta d} = 0.183 \left(\frac{75}{2.125}\right) = 6.45 \text{ m}^3/\text{hr}\cdot\text{m}$$

$$S = 2.246 \frac{Tt_o}{x^2} = \frac{2.246\,(6.45)\,(2.32)}{50^2} = 0.01344$$

2. If pumping continues at the same rate for 60 hours,

$$u = \frac{0.25\,x^2 S}{Tt}$$

$$= \frac{0.25\,(50)^2\,(0.07342)}{6.45 \times 60} = 2.45 \times 10^{-2}$$

From Table 4.1 the value of $f(u) = 3.281$.
Drawdown at the observation well after 60 hours,

$$d = \frac{Q}{4\pi T}\,f(u) = \frac{75 \times 3.281}{4 \times 3.14 \times 6.45} = 3.04\ m$$

3. If pumping continues at the same rate for 6 months

$$u = \frac{0.25 \times 50^2 \times 0.01342}{6.4 \times 6 \times 30 \times 24} = 3.075 \times 10^{-4}$$

From Table 4.1 the value of $f(u) = 7.512$

$\therefore$ Drawdown at the observation well after 6 months,

$$d = \frac{Q}{4\pi T}\,f(u) = \frac{75 \times 7.512}{4 \times 3.14 \times 6.45} = 6.95\ m$$

4.13 WELLS

Wells can be classified as shallow and deep. Wells up to 35 m depth are classified as shallow wells. A shallow well receives water from the subsoil overlying an impervious stratum. It may be contaminated by the surface water percolating through the soil from nearby areas.

A deep well (more than 35 m deep) receives water from an aquifer below an impervious strata. Chances of such a well becoming contaminated are remote.

A pipe sunk in the ground for drawing water is known as a tube well. The pipe can be driven or implanted in a drilled or bored hole.

4.14 TYPES OF WELL CONSTRUCTION

4.14.1 Dug Well

The oldest form of a well is the dug well, excavated by hand to shallow depths (Figure 4.15), and is generally used for individual houses. It is lined with cement jointed bricks or concrete. The wall in the lower part is made of bricks with open joints to allow water to percolate inside. Excavation is continued until water starts rushing into the well. Depths may vary between 5 and 15 m depending upon the level of the water table. The diameter of the

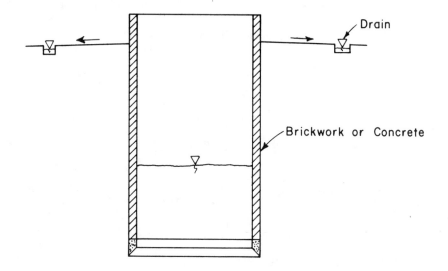

Figure 4.15. Dug well.

well may be between 1 and 5 m. Due to the large dimensions of the well, it also functions as a storage chamber. The top of the well may be open or closed depending upon how the water will be drawn from the well, but for sanitary reasons it should be covered. The well lining should be raised at least 0.3 m above the ground level and be plastered with grout. All surface water should drain away from the well, and the surface around the well should be covered with concrete.

4.14.2 Driven Well

A driven well is the simplest method of drawing water from shallow depths (Figure 4.16). The wells can be driven by a driving weight suspended through a pulley attached to a tripod. The steel pipe may be approximately 50 mm in diameter, with a strainer and a driving point or a cutting edge at the end. The driving point is slightly larger than the casing. Hand pumps are usually employed for drawing water, although pumps driven by electrical power may also be used. A suitable platform and good drainage must be provided all around the well.

4.14.3 Bored Well

Augers driven by hand or machine are used to construct bored wells. The soil should be cohesive so that when augers are taken out for cleaning and removing the soil, the sides of the hole do not cave in. The well casing is

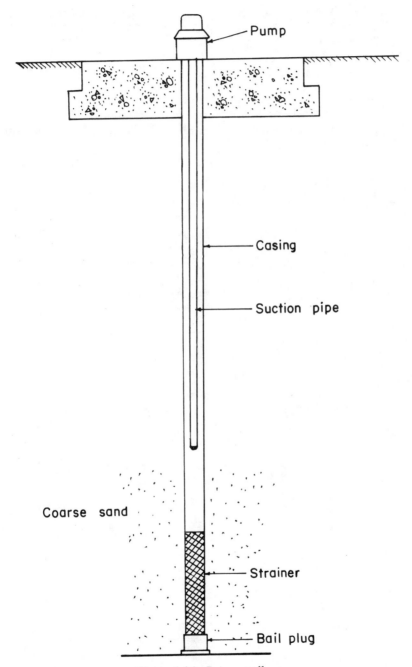

Figure 4.16. Driven well.

placed by driving after the auger has reached the water bearing stratum. The casing is cemented in place. The diameter of bored wells ranges between 250 mm and 600 mm.

4.14.4 Drilled Well

Most deep, high capacity wells are constructed by drilling. The diameter of drilled wells ranges between 150 mm and 1000 mm (Figure 4.17). The type of construction employed varies with local conditions, region, and with the driller.

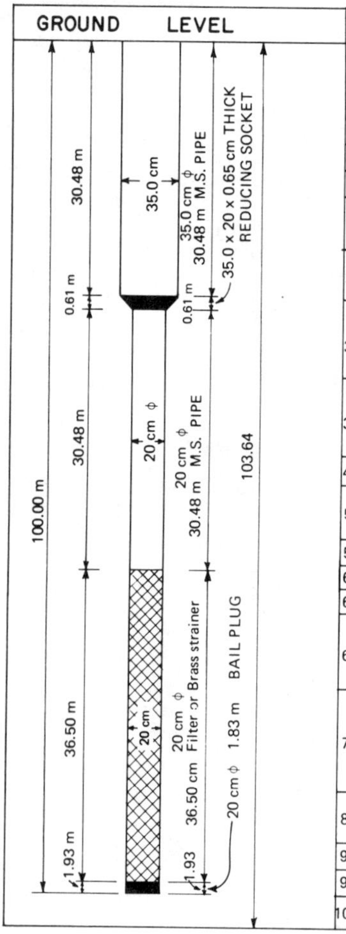

GROUND LEVEL	DEPTH from (m)	to (m)	THICKNESS (m)	SOIL STRATA
	0	6.09	6.09	RED CLAY
	6.09	9.14	3.05	CLAY LOAM
	9.14	2.19	3.05	SAND WITH CLAY
	12.19	15.24	3.05	FINE SAND
	15.24	18.29	3.05	MEDIUM SAND
	18.29	24.38	6.09	FINE SAND
	24.38	30.48	6.10	MEDIUM SAND
	30.48	39.62	9.14	FINE SAND
	39.62	48.77	9.15	MEDIUM SAND
	48.77	51.82	3.05	MEDIUM SAND
	51.82	57.91	6.09	FINE SAND
	57.91	60.96	3.05	MEDIUM SAND WITH STONE
	60.96	64.01	3.05	COARSE SAND WITH PEBBLE
	64.01	67.06	3.05	COARSE SAND
	67.06	76.20	9.14	COARSE SAND WITH PEBBLE
	76.20	88.38	12.19	COARSE SAND
	88.39	94.49	6.10	COARSE SAND WITH PEBBLE
	94.49	97.54	3.05	COARSE SAND
	97.54	100.59	3.05	COARSE SAND WITH PEBBLE
	100.59	103.64	3.05	COARSE SAND WITH STONE

Figure 4.17. 20 cm diameter deep drilled well with log chart.

4.15 METHODS OF DRILLING

The following methods are usually employed to drill wells:

a) Percussion drilling: soft clay to hardest rock
b) Hydraulic rotary: ⎫
c) Core drill method: ⎭   hard rock
d) Water jet boring: soft, unconsolidated alluvial deposits

4.15.1 Percussion Method

The drilling bit is raised and dropped to cut, break, and crush the soil. This is done repeatedly until the bit penetrates the strata. The powdered material is removed by jetting water under pressure. The cuttings are then removed by lowering a bailer into the hole.

4.15.2 Hydraulic Rotary Method

In this method a fluid consisting of a mud slurry is pumped into the bottom of the well through the stem of a hollow drill and a hole in the cutting tool. The slurry returns to the surface of the well, bringing with it the materials cut by the drill. In this method drilling proceeds rapidly and as boring proceeds the hole is lined to prevent caving of the sides.

4.15.3 Core Drill Method

Diamond drill bits are used for drilling holes in hard rock, or a hard hollow steel cylinder may be employed. The core drill rotates and cuts with the help of teeth provided on the annular circumference. Solid cores of rock are cut, which are taken out and preserved. These cores indicate geological formations.

4.15.4 Water Jet Boring

Water is forced through a jet at the lower end of the drill bit. The drilling proceeds by raising and lowering the drill bit, while water loosens the soil and brings with it finer portions of the loosened material to the top. The casing is usually lowered while boring proceeds.

4.16 CASING

The purpose of a well casing is to avoid the following:
1.   Caving in of the bore hole
2.   Entrance of unwanted water inside the well
3.   Seepage of good water from the well to the surrounding soil
4.   Mixing of soils from the sides with water

Depending upon the soil condition, the well casing may be left in place, partially removed, or withdrawn completely (Figure 4.18). It is not advisable to use the well casing as a part of the discharge pipe from the well.

The materials generally used for casings are wrought iron, alloyed or unalloyed steel, and cast iron. In some instances, copper pipes or asbestos cement pipes are used.

4.17 SCREENS

In a cased well, the casing is driven the full depth. After that the screen is lowered in the well until it rests on the bottom. The casing is then raised to completely expose the strainer (Figure 4.17) and the joint between the casing and the screen is sealed.

The operation and the life of a well depend to a great extent on the shape and size of the openings in the screen. The screen should not allow fine and medium sized sand to enter the well and at the same time it should offer little resistance to the flow of water.

Screen clogging can be minimized by using openings which are larger on the inside of the well (Figure 4.19). The materials used for screens are iron or

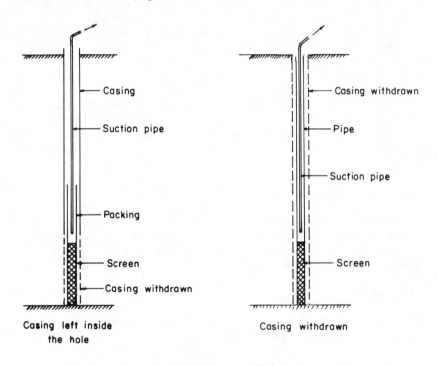

Figure 4.18. Well construction.

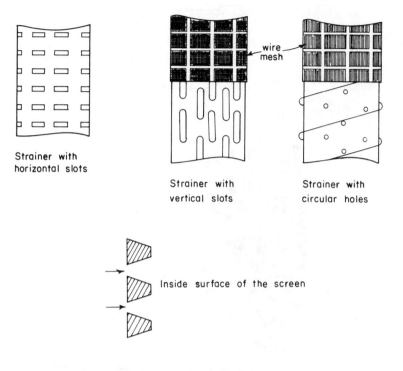

Strainer with
horizontal slots

Strainer with   Strainer with
vertical slots   circular holes

Inside surface of the screen

Section through the wall of a screen

Figure 4.19. Screens. (Courtesy of R. Oldenbourg Verlag GmbH.)

steel, corrosion resistant alloys, copper, or non-metallic substances like stoneware, wood, or asbestos cement.

4.18 GRAVEL WALL WELL

In a gravel wall well, the screen is covered by a gravel wall (Figure 4.20). This is generally used where the aquifer consists of very fine, uniform material. The size of the gravel is chosen so that the entrance of sand particles of unwanted size is prevented.

4.19 RANNEY WATER COLLECTOR (HORIZONTAL STRAINER WELL)

These wells yield large quantities of water (20,000 to 50,000 m³/day). They are especially suitable as infiltration works adjacent to the banks of rivers or lakes which contain coarse sand, gravel, and small pebbles. A Ranney water collector is a dug well of about 4 to 5 m in diameter and is sunk to hard stable strata (Figure 4.21). It is sealed at the bottom. Horizontal well screens

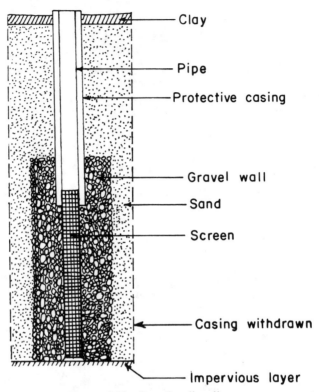

Figure 4.20. Gravel wall well.

are driven radially from the well, the length of which varies between 30 to 100 m. The maximum number of screens in a single well is twelve. Powerful hydraulic jacks (100 to 200 ton capacity) are used for driving the screens into the water bearing strata. Steel screens 200 mm in diameter having thick walls (8 mm) are generally used as horizontal feeders.

Horizontal feeder wells essentially consist of the following:

1.   Water tight caisson, which functions as a collecting well for water discharged from the horizontal feeders
2.   Horizontal feeders driven radially into the water bearing strata
3.   Sluice valves on the horizontal feeders with valve control, spindles and wheels
4.   Valve control platform
5.   Pumps and pump house

The drawdown curve for a horizontal strainer well is in the form of a depression cone with radial furrows (Figure 4.22) covering the horizontal feeder. As it is not possible to derive an exact expression for a horizontal strainer well, approximations are made to arrive at a formula. It is difficult because of many variables involved, i.e. the permeability, the thickness of the

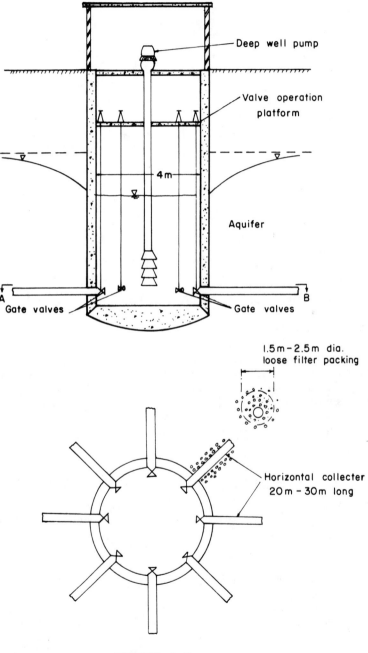

Deep well pump

Valve operation platform

4 m

Aquifer

A

B

Gate valves

Gate valves

1.5 m − 2.5 m dia.
loose filter packing

Horizontal collecter
20 m − 30 m long

SECTION A−B

Figure 4.21. Ranney water collector (after Busch[8]).

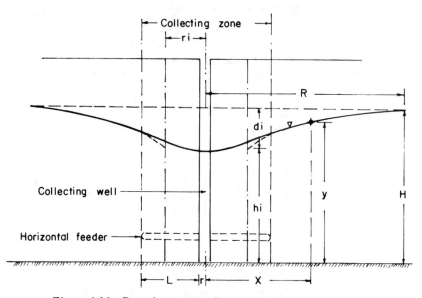

Figure 4.22. Drawdown curve for a horizontal strainer well.

water bearing strata, the drawdown, the length of feeders, the diameter and the number of feeders.

Pumping tests have indicated that the shape of a drawdown curve is similar to the drawdown curve of a vertical well. The equation of a drawdown curve of a horizontal strainer well can be calculated by the formula derived for an assumed vertical well, taking $x \geqslant L + r$ as the ideal radius r_i

$$r_i = \frac{A_i}{2\pi h_i} \quad \text{(for unconfined aquifer)} \dots \dots \dots (4.43)$$

and

$$r_i = \frac{A_i}{2\pi m} \quad \text{(for confined aquifer)} \dots \dots \dots (4.44)$$

in which

A_i = total surface area of the filter packing around the horizontal strainer

$h_i = H - d_i$ = height of the wetted screen surface of the assumed vertical well

m = the thickness of the confined layer

$$y^2 = \frac{Q \ln \frac{x}{r_i}}{\pi K} + h_i^2 \quad \text{(unconfined aquifer)} \dots \dots (4.45)$$

and $$y = \frac{Q \ln \frac{x}{r_i}}{2\pi m K} + h_i \quad \text{(confined aquifer)} \dots \dots (4.46)$$

Equations for drawdown curves for $x \geqslant L + r$

Consequently

$$Q \approx \frac{\pi K (H^2 - h_i^2)}{\ln \dfrac{R}{r_i}} \quad \ldots \ldots \ldots \ldots \ldots \ldots (4.47)$$

and

$$Q \approx \frac{2\pi m K (H - h_i)}{\ln \dfrac{R}{r_i}} \quad \ldots \ldots \ldots \ldots \ldots (4.48)$$

in which

R = 3000 $d_i \sqrt{K}$ and Q is the approximate yield of the well for unconfined and confined aquifer respectively.

These equations can be used only when at least three horizontal feeders are symmetrically placed.

The drawdown in the collecting zone $x < L + r$ cannot be determined due to the complicated flow phenomena in that zone. The ideal radius r_i can be approximated by the following expression:[7]

$$r_i = \frac{2}{3} \frac{\Sigma L}{n}. \quad \ldots \ldots \ldots \ldots \ldots \ldots (4.49)$$

4.20 YIELD FROM SHALLOW WELLS

The following two methods are employed for determining the yield of an ordinary shallow well in which the stream lines do not remain parallel but tend to converge upwards:

1. Pumping test
2. Recuperation test

4.20.1 Pumping Test

A pumping test is performed during the driest part of the year. The rate of pumping should be such that the level of water in the well remains steady. The volume of water pumped is then measured, and this will give the discharge.

4.20.2 Recuperation Test

First the level of water in the well is observed. The well is then pumped until the level of water drops to a particular level after which pumping is stopped, and the well refills. It has been observed that the discharge from the well is proportional to the depression head (H - d) (Figure 4.23).

$$Q = K (H - d) = Kh \quad \ldots \ldots \ldots \ldots \ldots (4.50)$$

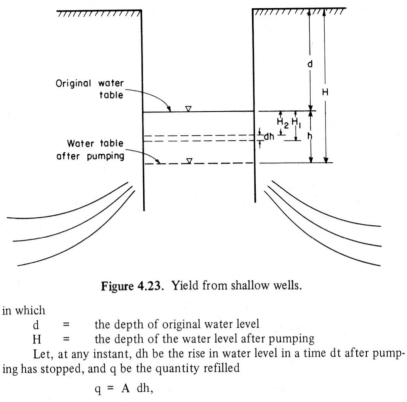

Figure 4.23. Yield from shallow wells.

in which
 d = the depth of original water level
 H = the depth of the water level after pumping
  Let, at any instant, dh be the rise in water level in a time dt after pumping has stopped, and q be the quantity refilled

$$q = A\ dh,$$

in which
 A = area of the well

Also $q = Q\ dt$

Therefore, $- A\ dh = Q\ dt = Kh\ dt$

$$- \frac{dh}{h} = K \frac{dt}{A}$$

  Integrating we get

$$- \int_{H_1}^{H_2} \frac{dh}{h} = \frac{K}{A} \int_0^t dt$$

$$\ln \frac{H_1}{H_2} = \frac{K}{A} t$$

$$\frac{K}{A} = \frac{2.303}{t} \log \frac{H_1}{H_2} \qquad \cdots \cdots \cdots \cdots \cdots (4.51)$$

in which
> H_1 and H_2 are the depths of water below original water level in time t
>   and K is the specific yield of the well

4.21 SPRINGS

In collecting water from springs care has to be taken not to disturb the natural soil conditions. The surface water near the spring should not be allowed to penetrate the soil strata and mix with the spring water.

A cut-off wall should be constructed fairly deep in the water bearing strata for collecting the water in a chamber (Figure 4.24). The collecting chamber should be provided with necessary pipelines, valves, and a manhole. The tank can be emptied by providing an overflow weir and a pipe. If water brings with it large quantities of sand then a pre-settling chamber should be constructed. On the surface, a diversion ditch around the collecting chamber should be provided for quick removal of surface water.

4.22 RECHARGING

The groundwater yield may be augmented by recharging. Apart from this, recharging may be utilized to purify the water. River water prior to recharging may be allowed to settle in a settling chamber having an impermeable bottom, or it may be allowed to pass through a rapid sand filter. After that, water is permitted to percolate through a selected bed of permeable soil. Charging tanks or ditches (Figure 4.25), vertical wells (Figure 4.26), infiltration galleries, and horizontal strainer wells may be suitably employed for recharging the underground water source. Geological conditions will indicate which methods should be used. Open tanks and ditches are suitable where the soil between the bottom of the tank and the groundwater table is sufficiently permeable. On an average the rate of seepage may be taken as 1 m^3 of water per m^2 of percolating surface per day. The sand layer at the bottom of the charging tank may be replaced after it is choked by the sludge in the water.

If the raw water contains excessive organic impurities or excessive iron, it should be treated prior to recharging, for this protects the filter of the wells. The recharged water can be withdrawn the same way the groundwater source is tapped. The horizontal distance between the recharging tank and the withdrawal plant will depend upon the permeability of soil. The distance may be based on a time of flow of 40 to 60 days. To find the quantity of water to be recharged, a pumping test in an observation well may be conducted. Instead of a drawdown curve, an infiltration cone at the recharging well is formed. Recharge rates between 1.5 m/day and 22.5 m/day have been achieved.

The equation for a recharging well can be formulated assuming that the infiltration is the image of the withdrawal.

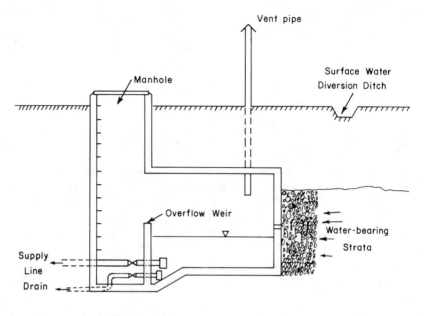

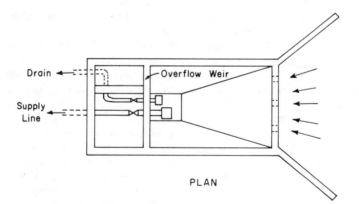

Figure 4.24. Collection of water from spring (after Brix-Heyd-Gerlach[10]).

For an unconfined aquifer,

$$- I = \frac{\pi K (H^2 - h^2)}{\ln \dfrac{R}{r}}$$

$$I = \frac{\pi K (h^2 - H^2)}{\ln \dfrac{R}{r}} \quad . \; . \; . \; . \; . \; . \; . \; . \; . \; . \; . \; . \; . \; (4.52)$$

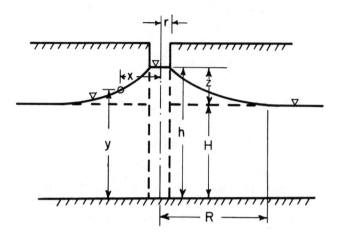

Figure 4.25. Recharging of well having free water table.

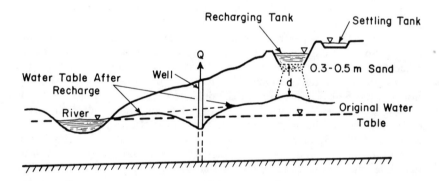

Figure 4.26. Groundwater recharge.

in which

    -I   =   percolation rate per unit of time (m^3/hr)

    h   =   H + Z = wetted length of the screen surface (m)

    Z   =   pressure of water on the circumference of the recharging well (m)

For a confined aquifer

$$I = \frac{2\pi mK\,(h - H)}{\ln \dfrac{R}{r}} \quad \ldots \ldots \ldots \ldots \ldots \quad (4.53)$$

For open tanks and ditches, the bottoms of which do not come into contact with the water table, the percolation rate can be calculated as follows:[8]

$$I = A v_a \qquad \dots \dots \dots \dots \dots \dots \dots \dots (4.54)$$

Approach velocity can be estimated as:

$$v_a \approx \frac{K (d + d_r)}{2d} \qquad \dots \dots \dots \dots \dots \dots (4.55)$$

in which

d =   distance between the water table after recharging and tank bottom

d_r =   depth of water in the recharging tank

K =   average permeability of the soil lying between the bottom of recharging tank and the water table after recharging

A =   soil surface area of the recharging tank

4.23 STORAGE CAPACITY OF RESERVOIRS

When the flow in the stream during dry weather is sufficient to meet the demand of a community there is no need for a storage reservoir for water supply, but when it is less than the demand, storage reservoirs need to be constructed.

The yield of a stream is close to the mean annual flow rate after the construction of a dam or a weir across it. Yield is affected by the loss due to evaporation, loss due to seepage, flow of water to and from the bank storage and the silting of the reservoir. To ensure a continuous withdrawal, the storage must be sufficient to meet the demand during dry periods. This can happen if the yield is equal or more than the demand. To determine the reservoir capacity, the average monthly flow in the stream over a reasonable period and the probable demand for water during that period should be known. At the beginning of the dry period it is assumed that the reservoir is full. After that gradually the reservoir starts to empty. At the end of the dry period there will be maximum deficiency. The storage capacity should be equal to the maximum deficiency, which can be expressed as follows:

$$S = \text{maximum deficiency} = \text{maximum } (D - Q) \qquad \dots \dots \dots (4.56)$$

in which

S =   storage capacity of the reservoir

D =   demand

Q =   flow in the stream

The value of S can be found arithmetically or graphically. With the help of a mass diagram, storage capacity can be conveniently found. To illustrate this, Figure 4.27 shows the cumulative stream runoff OA and the cumulative demand OB with respect to time. If a line parallel to OB is drawn tangential

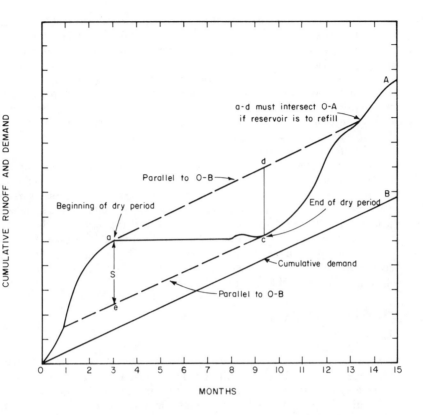

Figure 4.27. Mass diagram for determining the storage required in impounding reservoir.

to the point a, the starting point of the dry period, the ordinate cd will give the storage required to maintain a rate of demand represented by the line OB. A line drawn parallel to OB and tangent at c when projected must intersect the curve OA, if the reservoir is to be full at the beginning of the dry period.

While developing an impounded reservoir for collection of surface water, due consideration should be given to the site characteristics, the catchment area and preparation of the site.

4.24 SELECTION OF RESERVOIR SITE

The materials for the construction of the dam and the spillway should be easily available and in adequate supply. The surface and sub-surface geology should offer resistance against seepage of the impounded water. No objectionable mineral or salt which may affect the quality of water should be present.

Topography should be such as to have a narrow gorge so that the length of the dam is small. The valley on the upstream side of the dam should be broad so as to have a greater average volume per unit height and per unit length of the dam.

The quality of the runoff can be improved by utilizing the self purification process to its maximum. Factors which influence self purification are detention time, chemical and biological processes, sun light, temperature, wind action, and plankton life. The detention time is the most important, and 30 to 40 days is usually considered sufficient.

The reservoir should be shaped to minimize short circuiting incoming water to the intake point. Narrow reservoirs with their major axis along the direction of the prevailing wind may encourage short circuiting thus reducing detention time. The side slopes should be steep so as to have a small surface area per unit volume. This will reduce evaporation and undesirable effects of shallow water. Trees and bushes present inside the reservoir should be removed. Objectionable vegetation and marshy land should be absent.

Important roads and railway lines should be avoided as they may become submerged. Human habitation should be affected to a minimum and submergence of cultivable lands is to be avoided.

4.25 CATCHMENT AREA

When impounded reservoirs have to be developed in valleys and between hills solely for water supply purposes, then the entire watershed should be acquired. Human habitation must be removed from the area to ensure a supply of safe water. As far as possible, cultivable lands should be kept out of the watershed. This keeps the water free of turbidity and nutrients.

There must be proper control of the wastewater discharge. Proper drainage of the swamps and marshy areas must also be considered. Soil erosion should be minimized. Where water is to be supplied from large lakes and rivers, the discharge of wastewater in the water courses must be properly controlled. In this case, water should be supplied to the community only after suitable treatment. But much can be achieved by regulating the wastewater discharge and relying on the self purification capacity of the water courses, especially in terms of the raw water quality and the water treatment cost.

4.26 PREPARATION OF SITE

Rehabilitation is an important problem. Because there is a psychological problem for those who have to leave their ancestral homes, every possible care should be taken in handling this problem. The displaced people should be taken into confidence so that they understand the importance of the work, and know it is in their interest, as well as in the interest of the nation. The family members should be provided with employment. Houses and other facilities should be quickly provided. Goodwill and understanding should be developed in the area.

Vegetation is removed by cutting down and removing the trees, bushes, etc. Marshy areas are either drained out or filled in. Highly organic surface soil like peat and cultivable lands may have to be removed. This will reduce the undesirable effect of decaying organic matter.

4.27 WATER INTAKE

The following points should be considered in the location of an intake:

1. The intake should be located in a place where there is no fast current which may damage the intake causing interruption in the water supply.

2. The ground near the intake should be stable. A straight section of the river is always preferable as the risk of erosion of the bank in this case is minimum.

3. The approach to the intake should be free from obstacles.

4. The inlet should be well below the surface of the river or the lake for receiving cooler water and for preventing the entry of floating matter. To prevent the entry of suspended matter near the bottom, the inlet point should also be well above the bottom of the water body.

5. To avoid possible contamination of the bank, the intake should be located at some distance from the bank.

6. Intake should be located on the upstream of the town.

The main current, lowest water level in the river, and navigation facilities should be considered in locating the intake. Curtain walls may be constructed for diverting water into intake structures at the bank in order to draw clear water from the stream. Intake structures and coarse and fine screens should be constructed where they will not be damaged by floods. They should also be safe from scouring or silt deposition. Coarse as well as fine screens should be provided so that floating matter may not enter the supply system. Considering the fluctuation in water level, inlets are provided in the well at various levels (Figure 4.28). If the fluctuation in the water level between the summer and rainy season is too large and the river becomes almost dry in summer, water should be stored by constructing a small weir across the river. By so doing, compensation can be made for a shortage of water in the river during the dry period.

When the level of water in the river is fairly constant and the river bank is steep, the intake works can be constructed adjacent to the bank. In this case water is drawn from the river through a pipe laid horizontally (Figure 4.29). The inlet is provided with a coarse screen and strainers and is firmly anchored (Figure 4.30).

4.28 DIRECT INTAKE (Figure 4.31)

Direct intakes are located in deep waters. It is cheaper than any other type of intake.

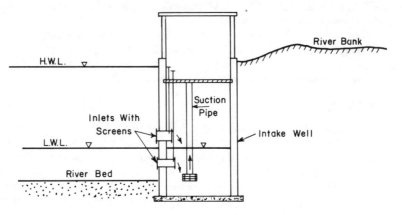

Figure 4.28. River intake.

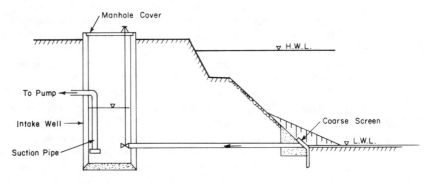

Figure 4.29. River intake.

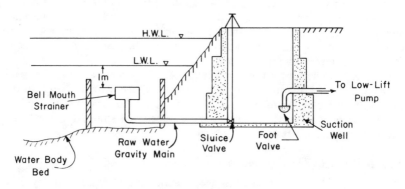

Figure 4.30. River intake.

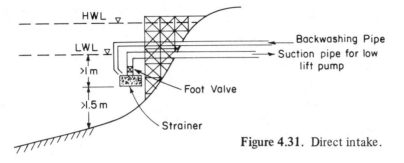

Figure 4.31. Direct intake.

Under the following conditions it may be used suitably:
1.   When the source is deep such as rivers, lakes.
2.   When the embankment is resistant to erosion and sedimentation.

4.29 CANAL INTAKE

When water is drawn from a canal, a masonry chamber with an opening is built partially in the canal bank. The opening is provided with a coarse screen (Figure 4.32). From the chamber, water is drawn by a pipe having a bell mouth covered with a perforated hemispherical cover. The area of the hole in the cover is one-third the area of the hemisphere.

Due to the construction of the chamber in the canal, the width of the canal is reduced causing an increase in the velocity. This may scour the soil and therefore the approaches in the upstream and downstream are lined with rip-rap.

4.30 RESERVOIR INTAKE

The intake tower is either located at the spillway section or near the toe of an earthen dam (Figure 4.33). The foundation of the tower is separated from that of the dam. It is constructed on the upstream side. A number of inlets at various levels are provided by the tower to compensate for water level fluctuations. When reservoirs are at such a level that water can flow by gravity to the water purification works, then an intake tower is not required (Figure 4.34).

4.31 ELEMENTS OF AN INTAKE

1.   Bell mouth strainer or cylindrical strainer (Figure 4.32).
2.   Strainer structure with arrangements for its protection
3.   Raw water gravity pipe or channel
4.   Gate or sluice-valve
5.   Suction well (intake well)
6.   Foot valve
7.   Suction pipe for the low lift pipe

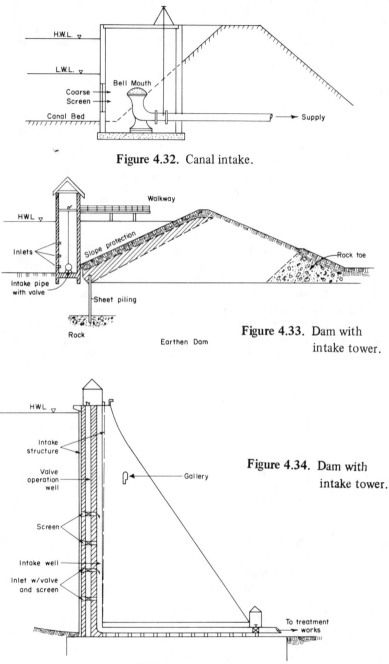

Figure 4.32. Canal intake.

Figure 4.33. Dam with
intake tower.

Figure 4.34. Dam with
intake tower.

Design Criteria

a) Bell mouth strainer:
   1. Velocity through the strainer hole = 0.15 to 0.30 m/s (0.5 to 1 ft/s). It is recommended that the velocity be near the lower limit to prevent the entry of impurities.
   2. Openings of the strainer holes = 6 to 12 mm (¼ to ½") diameter.
   3. The gross area of the strainer = 2 times the effective area, i.e. the total area of the holes.

b) Cylindrical strainer:
   The same criteria for design are recommended as those for the bell mouth strainer. A cylindrical strainer should be used when there is a high head of water above the strainer. The strainer should be 0.6 to 1 m (2 to 3 ft) below the lowest water level, if it does not have holes at the top. Strainers having holes at the top should be more than 1 m (3 ft) below the lowest water level.

c) Raw water gravity pipe:
   1. To prevent sedimentation and erosion, the velocity of water should be between 0.6 and 1.5 m/s (2 to 5 ft/s).
   2. The pipe is sized such that the velocity at the LWL will be greater than 0.6 m/s (2 ft/s) and at the HWL the velocity will be less than 1.5 m/s (5 ft/s). Knowing the head and the velocity, the proper diameter of pipe can be selected.

d) Suction well (intake well):
   1. From the maintenance point of view, there should be at least two wells.
   2. Detention time should be at least 20 minutes, or the well must be large enough to enter for cleaning.
   3. The bottom of the well should be at least 1 m (3 ft) below the river bed or 1.52 m (5 ft) below the lowest water level.
   4. The height of the foot valve above the bottom of the well should not be less than 0.60 m (2 ft).
   5. The well should be water-tight and constructed of durable material such as reinforced concrete. The wall thickness should be 20 cm (8 in) or thicker.
   6. The well should be heavy enough to withstand the uplift pressure.

e) Suction pipe of low lift pump:
   1. The velocity in the pipe should be between 1 m/s and 1.5 m/s (3 ft/s and 5 ft/s).
   2. The difference in height between the lowest water level and the center of the pump should not be more than 3.7 m (12 ft).
   3. When the level of the pump is higher than the LWL, then the suction distance should be less than 4 m (13 ft).
   4. A pump located below the LWL with a flooded suction line is preferable, whenever justified economically.

f) Backwashing pipe for cleaning foot valve and strainer:
   1. Velocity in the pipe should not be less than 3 m/s (10 ft/s).

2.   Treated water should be used.
3.   The quantity of backwash water should equal 1/3 of the flow in
the suction pipe.

Example 4.8

Design the suction pipe and the backwashing pipe for a flow of 3,790 ℓ/min
(1,000 gpm).
Assume the velocity of water through the suction pipe = 1.5 m/s (5 ft/s).
The cross sectional area of the suction pipe = 3.790/(1.5x60) = 0.042 m²
(0.393 ft²), or the diameter of the suction pipe is 0.23 m (0.7 ft or 9
in).
Select a pipe of standard diameter which is equal to or larger than the
calculated value.
Quantity of water for backwashing - 1/3 Q - 1/3 x 3,790 = 1,263.33 ℓ/min
(333.33 gpm).
Assume the velocity of water in the backwashing pipe = 3.05 m/s (10 ft/s).
The cross sectional area of pipe = 1,263.33/(3.05x60) = 0.007 m² (0.074
ft²).
Diameter of pipe = $(0.007)^{1/2}/0.785$ = 0.094 m (0.31 ft = 4 in).

Example 4.9

Design a strainer for a flow of 3,790 ℓ/min (1,000 gpm).
Assume the velocity through the strainer = 0.15 m/s (0.5 ft/s).

$$Q = AV \text{ where A is the effective area, V is the velocity of water}$$

$$A = \frac{Q}{A} = \frac{3.79}{0.15 \times 60} = \frac{3.79}{9} = 0.42 \text{ m}^2 \ (4.52 \text{ ft}^2)$$

Gross area = 2A = 2 x 0.42 = 0.84 m² (9.04 ft²)
For the bell mouth strainer

$$\text{diameter d} = \left(\frac{0.89}{0.785}\right)^{1/2} = 1.034 \text{ m} \ (3.39 \text{ ft})$$

For the cylindrical type of strainer without holes on the top and assuming a
height of 0.61 m (2 ft) the perimeter and diameter are calculated as follows:

$$\text{Perimeter} = \frac{0.89}{0.61} = 1.312 \text{ m} \ (4.52 \text{ ft})$$

$$\text{Diameter d} = \frac{1.312}{3.14} = 0.42 \text{ m} \ (1.43 \text{ ft})$$

Example 4.10

Design a suction well (intake well) for a flow of 3,790 ℓ/min (1,000 gpm) if
the highest water level and the lowest water level are 9 m (30 ft) and
4.27 m (14 ft) above the river bed, respectively. Ground level is 9.76 m
(32 ft) above the river bed.

Assume the detention time is 20 minutes.

Volume of the well = 3,790 x 20 = 75,800 ℓ (2,683 ft^3).

Choose two suction wells.

Volume of each suction well = 75.8/2 = 37.9 m^3 (1,342 ft^3).

The bottom of the well will be located 1.5 m (5 ft) below the lowest water
level.

Effective depth of the well = 9 - (4.27-1.5) = 6.23 m (20.42 ft).

Assume a free board of 0.61 m (2 ft).

Hence the total depth of the well = 6.23 + 0.61 = 6.84 m (22.42 ft).

Therefore the area = 37.9/6.84 = 5.54 m^2 (59.85 ft^2).

Using a circular well, d = (5.54/0.785)$^{1/2}$ = 2.66 m (8.73 ft).

5. Distribution of Water

5.1 METHODS

Depending upon the topography, the location of the source, and other considerations, water can be transported to a community in a number of ways. For transportation of water, canals, flumes, tunnels, and pressure pipes can be employed. Water can be supplied to the consumers either by means of gravity, or pumping or a combination of both.

5.1.1 Gravitational Flow

When the source of water is a lake or an impounded reservoir at such a height that the desired pressure can be available at the consumer end, then a gravitational flow method of distribution can be employed. This is the most satisfactory means of water distribution.

5.1.2 Direct Pumping

Water is pumped directly into the mains. This method has several disadvantages. In case of a power interruption the water supply comes to a standstill. The variation in water consumption affects the pressure in the mains. To compensate for this a number of pumps of varying capacities are installed and they have to be put in operation according to the water requirements. This method needs careful operation and good maintenance.

5.1.3 Pumping with Storage

This method is very reliable. The excess water during low consumption is stored in elevated tanks, and utilized during peak hours. This method is economical as the pumps can operate at their rated capacity, and chances of damage to pipe appurtenances due to pressure variation is reduced. The stored water in the elevated tank forms a reserve for any eventuality like fire or power failure. The minimum pressure in the pipeline even at the highest point in the area can be guaranteed.

5.2 SERVICE RESERVOIRS (DISTRIBUTION RESERVOIRS)

Service reservoirs may be constructed as tanks or water towers. They are used for equalizing the rate of flow and for emergencies. Depending upon the utility, purpose, and topography, reservoirs may be underground or elevated.

Elevated reservoirs may be made of earth or masonry or concrete. Water towers are usually made of steel or concrete. There can be a number of them in a city on elevated ground. They usually have a capacity between 100 and 1,000 m³.[1]   Generally the equalizing volume of a water tank is about 1/6 to 1/3 of the total daily demand. The equalizing volume and the storage required for a reservoir can be determined analytically or graphically. The total volume (equalizing plus storage) of the reservoir should be one day's demand.

5.3 DETERMINATION OF CAPACITY

5.3.1 Analytical Method

The draft expressed as percentage of maximum daily requirement is shown in Table 5.1 (column 2). The flow is confined to 10 hours, i.e., from 8 A.M. to 6 P.M. It is also expressed as percentage of maximum daily requirement (column 3). Cumulative storage is calculated and entered in column 4 (column 4 = column 3 - column 2). The capacity of the tank will be the sum of the maximum excess in the morning hours and the minimum deficiency in the afternoon (column 5). Storage required = (13) + (26) = 39 percent of maximum daily requirement. In addition to this, provision for fire fighting and leakage in the pipe should be made. The quantity of water for leakage may be taken as 10 percent of the sum of the water consumption and fire demand.

5.3.2 Graphical Method

(a) Calculate the cumulative draft from the values given in column 2 of Table 5.1. Plot these values (Figure 5.1) versus time.

(b) Draw a straight line from the beginning of the pumping period to the end of it, showing the supply or inflow. The sum of the maximum deviations indicated by ordinates y_1 and y_2 between the draft and inflow curves gives the storage.

The quantity of water in the service reservoir at any time during the day can be determined by plotting the ordinates between the draft and the shifted inflow line (Figure 5.1b). Thus the level of water in the reservoir can be found.

Table 5.1. Storage Capacity of a Service Reservoir

Time (Hours)	Draft in % of Max. Daily Requirement	Inflow in % of Max. Daily Requirement	Inflow-Draft %	Storage Variation in % of Max. Daily Requirement
(1)	(2)	(3)	(4) = (3) - (2)	(5) = Σ(4)
0-1	2	--	-2	- 2
1-2	1	--	-1	- 3
2-3	0.5	--	-0.5	- 3.5
3-4	0.5	--	-0.5	- 4.0
4-5	0.5	--	-0.5	- 4.5
5-6	2.5	--	-2.5	- 7.0
6-7	3	--	-3	-10.0
7-8	3	--	-3	-13.0
8-9	4	10	+6	- 7.0
9-10	6	10	+4	- 3.0
10-11	4	10	+6	+ 3.0
11-12	7.5	10	+2.5	+ 5.5
12-13	11	10	-1	+ 4.5
13-14	8.5	10	+1.5	+ 6.0
14-15	7	10	+3.0	+ 9.0
15-16	5	10	+5.0	+14.0
16-17	3	10	+7.0	+21.0
17-18	5	10	+5.0	+26.0
18-19	5	--	-5.0	+21.0
19-20	5	--	-5.0	+16.0
20-21	7	--	-7.0	+ 9.0
21-22	5	--	-5.0	+ 4.0
22-23	2	--	-2.0	+ 2.0
23-24	2	--	-2.0	0.0
Sum	100	100	0	--

5.4 SHAPE AND VOLUME OF WATER STORAGE FACILITIES

Water towers can be made of concrete or steel. These can be of various forms. The most suitable form for concrete towers is a cylinder with an intze shaped bottom or with a flat bottom (Figure 5.2a,b). Steel tanks may have a spherical or dome shaped bottom (Figure 5.2c,d).

The lowest water level in the tank is determined according to the pressure requirements in the pipeline. The pressure in the pipelines may vary

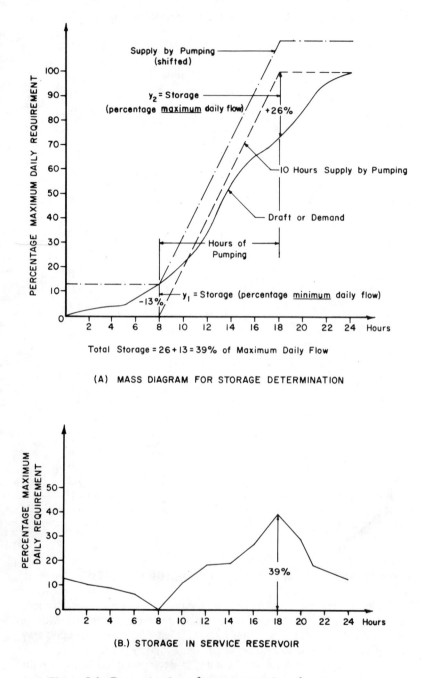

(A)   MASS DIAGRAM FOR STORAGE DETERMINATION

(B.) STORAGE IN SERVICE RESERVOIR

Figure 5.1. Determination of storage capacity of service reservoir.

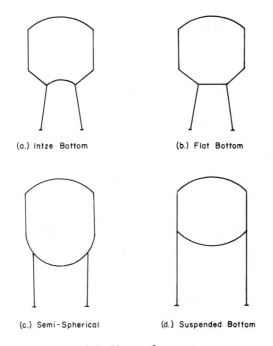

(a.) Intze Bottom   (b.) Flat Bottom

(c.) Semi-Spherical   (d.) Suspended Bottom

Figure 5.2. Shape of water towers.

from 15 to 50 m of water depending upon the type of community and the pressure needs of different areas in a city. The business zone of the city may need a high pressure. To keep the pumping cost low, the depth of water in the tank is generally kept small. Due to structural considerations the depth is kept equal to the diameter. For circular tanks the following approximate formula for the diameter may be used.[2]

Tank Volume V (m^3)	Diameter of the Tank (m)
100 - 500	$5 + \dfrac{V}{100}$
500 - 1000	$8 + \dfrac{V}{250}$

5.5 LOCATION OF SERVICE RESERVOIRS

A service reservoir stores the water and supplies it at the required pressure to the farthest point in the area. In view of the cost of pipelines and uniform distribution of pressure, the reservoir should be located near the center of the area (Figure 5.3). In flat areas this may be achieved by constructing a water tower at the center, but in undulated areas it may be more advantageous to select the highest point for the construction of an elevated tank, and this may be at one end of the area instead of at the center. Apart from the center, there are two possible locations for the elevated tank. It can be situated either between the area and the source of supply (pumping or gravity flow) or between the source and the elevated tank.

When the service reservoir lies between the area and the source, all the water must pass through the elevated tank before flowing through the area (Figure 5.3). Therefore, the pressure in the water supply system depends upon the water level in the service reservoir. In a water supply system a minimum pressure must be guaranteed even at the remotest point in the area; therefore, it is essential that the hydraulic gradient line always be above the required pressure (Figure 5.3a,b).

When water is supplied from an impounded high level reservoir (Figure 5.3b), the service reservoir also serves to reduce the high pressure of water. This reduces the possibility of damage to the pipe due to high hydrostatic pressure. Considering the strength of the pipeline, it is advisable that the difference in highest water level at the source and the lowest laid pipeline not be more than 60 to 70 m. If the difference in elevation is more, then the city may be supplied through a separate but commonly interconnected supply system. Areas having higher elevations are located in the high service system and areas of lower elevations are located in the low service system.

When the area lies between the source and the service reservoir, then most of the requirements are met by direct pumping (Figure 5.3c,d) and the excess water flows to the service reservoir. In this system there may be larger fluctuations in the supply pressure.

Example 5.2

The total water consumption during various hours in a city is given on page 78. The rate of pumping between midnight and 0600 hours and between 1800 hours and midnight is 50 percent of the average water consumption. During the rest of the day the rate of pumping is 150 percent of the average water consumption. Calculate the capacity of the service reservoir.

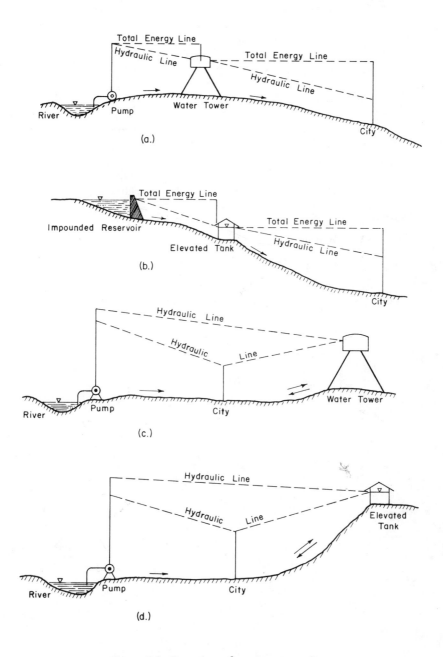

Figure 5.3. Location of service reservoir.

Time, hrs	Cumulative Water Consumption m^3
2	200
4	600
6	1200
8	2000
10	2900
12	3700
14	4500
16	5300
18	6000
20	6500
22	6900
24	7200

Solution:

Average water consumption = $\dfrac{7200}{24}$ = 300 m^3/hr

50% of the average = 150 m^3/hr
150% of the average = 450 m^3/hr

From Figure 5.4 we get the required storage = 300 + 300 = 600 m^3.

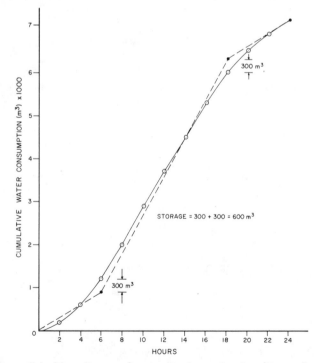

Figure 5.4. Mass diagram for storage determination (Example 5.2).

6. Pumping

6.1 GENERAL

The transport of water from low lying sources, e.g. underground water, rivers and lakes, to the elevated water towers, reservoirs, or directly to the consumers under pressure is accomplished with the help of pumps.

In the design of pumping works, stand-by units must be provided so that in case of breakdown or during repairs the water supply is not effected. The number of units in reserve will depend upon the particular situation and operational conditions. To achieve a higher degree of safety it may be advisable to have stand-by units operated by a different power source, e.g. the regular units run by electrical power and the reserve by diesel. There should be sufficient space in the pumping works to allow repair work to proceed easily. Space for installing an additional pumping unit should also be available. For proper maintenance and operation each unit should not be operated for more than 22 consecutive hours.

The pump house should be as close to the intake point as possible so that the length of suction pipe is short. The suction pipe should be air tight, but the water supply line has only to be water tight.

In pumping works various types of fittings like gate valves, check valves, water meters, and electrical control equipment are used. A gate valve must be installed behind the pump on the delivery side. If the pump is below the level of water, then a gate valve should be installed in the suction pipe prior to the pump. A check valve between the pump and the gate valve on the delivery side is installed to safeguard the pump from damage in case of breakdown.

Work done per second by a pump may be expressed as follows:

$$P'' = DgQH \, (\text{Watt}) \quad \ldots \ldots \ldots \ldots \ldots \ldots \quad (6.1)$$

For metric units

$$P'' = \frac{DgQH}{75} \quad (\text{mph}) \quad \ldots \ldots \ldots \ldots \ldots \quad (6.2)$$

in which
D = density of fluid (kg/m^3)
g = acceleration due to gravity (m/sec^2)
Q = rate of water flow (m^3/sec)
H = manometric head or total head (m)

$$\text{Power required } P = \frac{P''}{e_p} \quad \ldots \ldots \ldots \ldots \quad (6.3)$$

e_p =   efficiency of the pump (for reciprocating pump

$$e_p = 0.7 \text{ to } 0.95,$$

for centrifugal pump

$$e_p = 0.4 \text{ to } 0.85)$$

Taking into consideration the efficiency of power transmission from the electric motor to the pump, power will be

$$P'_M = P''/e_p e_{pt} \quad \ldots \ldots \ldots \ldots \ldots \ldots \quad (6.4)$$

in which

e_{pt} =   efficiency of power transmission (Table 6.1)

If the motor and the pump are directly coupled, then $e_{pt} = 1$.

If e_m is the efficiency of the motor, then the electrical power required will be

$$P_M = \frac{P'_M}{e_m} = \frac{DgQH}{e_p\, e_{pt}\, e_m} \text{ (Watt)} \quad \ldots \ldots \ldots \quad (6.5)$$

Total head H (gross lift) is the sum of the difference in the water level on the suction and discharge side and the head losses (Figure 6.1).

$$H = h + h_f = h_s + h_{fs} + h_d + h_{fd} \quad \ldots \ldots \ldots \quad (6.6)$$

in which

h_s =   suction head (m)
h_{fs} =   head losses on the suction side (m)
h_d =   height to which water has been raised above the pump (m)
h_{fd} =   head losses on the delivery side (m)

The velocity head $h_v = V^2/2g$ being small compared to other values may be neglected.

Table 6.1. Values of Efficiency (e_m) for Various Machines

Types	Efficiency e_m
Water turbine	0.75 to 0.85
Steam turbine	0.40 to 0.55
Gas motor	0.28 to 0.29
Diesel motor	0.34 to 0.35
Electric motor	0.85 to 0.91

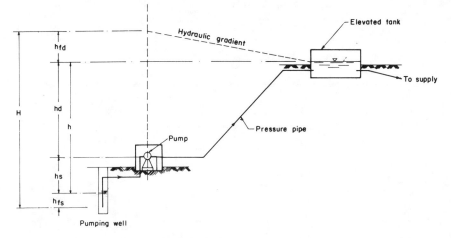

Figure 6.1. Calculation of total head.

6.2 SUCTION PIPE

The head which can be imparted to the water in the discharge pipeline depends upon the pump and the strength of the pipeline. But the length of the suction pipe is limited by the maximum height to which water can be lifted in the pipe. Although water is under atmospheric pressure (pressure head = 10.33 m of water), the height to which water may be raised is limited to 5 to 7 m of water. To calculate the allowable suction head, one has to consider the atmospheric pressure, the head loss in the suction pipe, the velocity in the suction pipe, and the saturated vapor pressure.

The vapor pressure depends upon temperature (Table 6.2). When hot water is to be pumped, the suction head may become too small or even negative due to the high value of vapor pressure. In this case water has either to flow under gravity or under pressure.

When the difference between the atmospheric pressure and the suction pressure is smaller than the saturated vapor pressure, the pipeline and the pump are damaged due to cavitation. The water is converted into vapor which strikes with a high velocity causing the material through which it flows to erode.

Example 6.1

Calculate the allowable suction head in a centrifugal pump when given the following information:

Table 6.2. Vapor Pressure Values at Various Temperature

Temperature, °C	Saturated Vapor Pressure (m of water)
0	0.063
4	0.083
10	0.125
20	0.238
30	0.433
40	0.752
50	1.258
60	2.031
70	3.177
80	4.83
90	7.15
100	10.33

Atmospheric pressure 730 mm mercury (760 mm mercury at $0°C$ = 10.33 m water)

Head losses in the suction pipe (friction losses, losses in fittings etc.) = 295 m

Velocity in the suction pipe = 1.6 m/sec

Temperature of water = 30° C

Vapor pressure = 0.433 m of water

Atmospheric pressure = 10.33 x 730/760 = 9.91 m of water

Velocity head = $v^2/2g$ = $(1.6)^2/2$ x 9.81 = 0.130 m of water

Theoretical lift = atmospheric pressure = 9.91 m of water

Sum of losses = 2.95 + 0.130 + 0.433 = 3.513 m of water

Allowable section lift = 9.91 - 3.513 = 6.397 m of water

6.3. PUMPS AND PUMPING

There are various factors to consider when pumping. These are: volume of water, pumping head, variation in flow, and type of power available.

The classification of pumps according to the mechanical principles on which they operate is as follows:

1.   Reciprocating pump
2.   Hand pump
3.   Centrifugal pump
4.   Air lift pump

6.3.1 Reciprocating Pump

A reciprocating pump consists of a cylinder in which a piston or a plunger moves backwards and forwards. Due to backward movement vacuum is created on the suction side and the forward movement creates pressure in the cylinder forcing the water to rise. Reciprocating pumps are of three types: (1) Single acting, (2) Double acting, and (3) Differential plunger.

The plunger, in a single acting reciprocating pump, will have acceleration at the beginning of the stroke and retardation at the end. As a result, there will be pressure variation in the cylinder and the flow will be pulsating (Figure 6.2). By fitting air vessels both on the delivery side and suction side, a constant flow may be maintained. The acceleration head is considerably reduced by fitting the air vessels as close to the cylinder as possible.

In a double acting reciprocating pump a more uniform flow, compared to a single acting pump, is achieved (Figure 6.3). It consists of two cylinders with one plunger. During the reciprocating motion of the piston, it creates pressure in one cylinder, while it produces a vacuum in another.

For overcoming greater heads, differential plunger pumps may be used. In this type the suction side is single acting and the delivery side is double acting (Figure 6.4). As the pump chamber and the discharge pipe are filled with water, the discharge valve closes with the backward movement of the plunger. During the forward motion of the plunger, the suction valve is closed and the water is forced out of the pump chamber through the delivery valve.

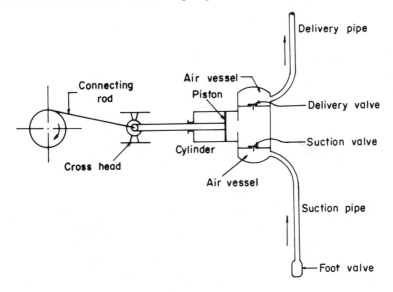

Figure 6.2. Reciprocating pump (single acting) with air vessel. (Courtesy of TH-Dresden.)

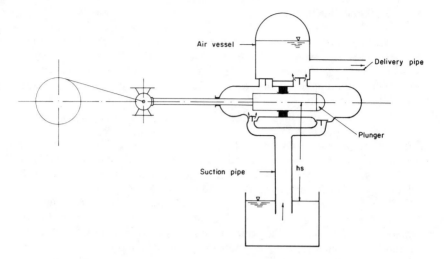

Figure 6.3. Double acting reciprocating pump. (Courtesy of TH-Dresden.)

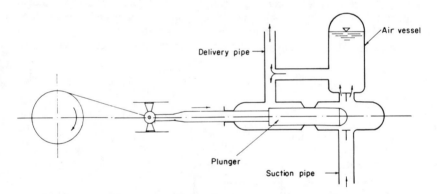

Figure 6.4. Differential pump. (Courtesy of TH-Dresden.)

The pressure in the pump must not fall below the separation pressure of the water, and therefore, the maximum speed of the pump is restricted by the acceleration head. The speed of the pump is seldom more than 100 rpm. The power requirement at the start is high, since the head against which the pump has to work and the quantity of water to be discharged remains constant throughout. The differential plunger pump is considered most efficient against high heads (over 100 m) and for large volumes of water. The disadvantages of this type of pump are the cost, the large space occupied and the difficulty of pumping sandy water.

6.3.2 Hand Pump

A hand pump is used to raise water from shallow depths. It is used most frequently for individual houses (Figure 6.5). It can raise water up to a height of about 7 m. This pump is single acting. During the upward motion of the piston the water is discharged as well as drawn in the cylinder through the valve in the suction pipe.

A vane type hand pump is frequently used to raise small quantities of water from shallow depths. The pump consists of an impeller mounted on a central shaft which can be given an oscillatory motion (Figure 6.6).

6.3.3 Centrifugal Pumps

The impeller of the centrifugal pump imparts to the water a centrifugal force, which creates a pressure head that raises the water to a certain height. The speed of the impeller and the size of its diameter determine the head to

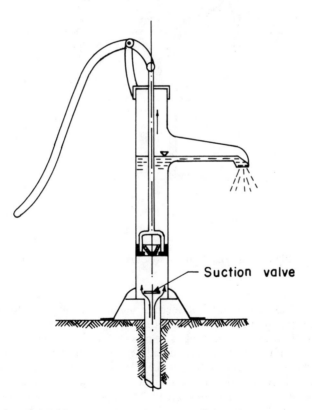

Figure 6.5. Hand pump. (Courtesy of TH-Dresden.)

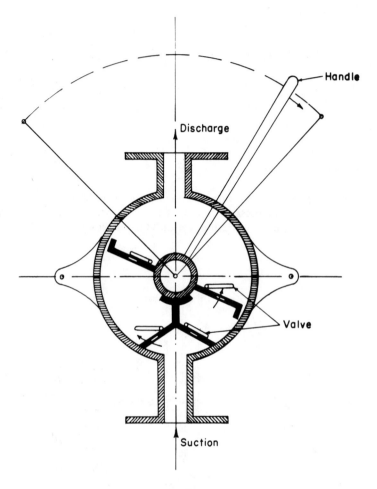

Figure 6.6. Hand pump. (Courtesy of TH-Dresden.)

which water can be pumped. The speed of the pump is generally 900 to 3,000 rpm. Small pumps are designed for speeds up to 6,000 rpm. The pump must be filled with water while starting. This process is called priming.

In centrifugal pumps, water usually enters at the center and flows out through the impeller radially (Figure 6.7). The kinetic energy of water leaving the impeller is converted into pressure energy by allowing the water to flow through a spiral shaped chamber (Figure 6.8). This can be achieved by making the water flow through passages of increased area provided by fixed guide vanes or diffusion vanes (Figure 6.9). This type pump has some resemblance

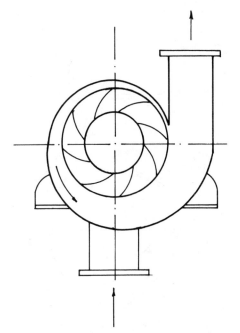

Figure 6.7. Centrifugal pump with volute chamber. (Courtesy of Pitman
Publishing Limited.)

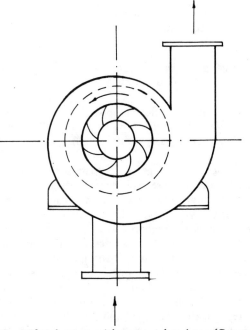

Figure 6.8. Centrifugal pump with vortex chamber. (Courtesy of Pitman
Publishing Limited.)

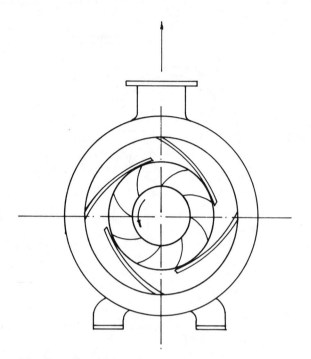

Figure 6.9. Centrifugal pump with guide blades. (Courtesy of Pitman
Publishing Limited.)

to a reaction turbine, and it is sometimes confusingly called a turbine pump.
In centrifugal pumps the flow through the impeller is radially outward
(Figure 6.10a).

Mixed flow or diagonal flow pumps represent an intermediate stage
between the axial and the radial type (Figure 6.10b). Axial flow pumps,
known as propeller or screw pumps (Figure 6.10c), are generally used for low
heads.

Pumps may be single stage or multistage. In a multistage pump a
number of impellers are put in series in a single casing, so that discharge from
one impeller is sucked into the next one. The quantity of water discharged
from a multistage is the same as that from a single stage, but the total head
developed is the product of the head of the single stage and the number of
stages. In a single stage pump the head developed can be as high as 100 m.
The maximum number of stages in a centrifugal pump can be 15 to 20. Thus
a multistage pump can raise water to any desired practical head.

A special type of centrifugal pump is the deep well pump (Figure 6.11).
The pump and the motor are assembled as one unit. The outer diameter of
the pump may be as small as 100 mm, and therefore, the impeller diameter is
smaller. Deep well pumps are multistage and because the diameter of the

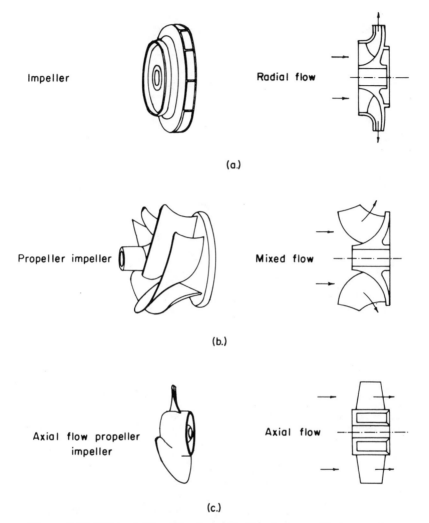

Impeller

Radial flow

(a.)

Propeller impeller

Mixed flow

(b.)

Axial flow propeller
impeller

Axial flow

(c.)

Figure 6.10. Shapes of impeller in a centrifugal pump. (Courtesy of
R. Oldenbourg Verlag GmbH.)

impeller is small, a single stage pump may not lift water to the desired height.
The pump is installed in a well casing, about 2 to 3 m below the lowest water
level, so that the suction pipe is eliminated. It is also known as a submersible
pump.

Irregardless of the lower efficiency and smaller suction head, the
centrifugal pump is most commonly used for pumping water. The advantages
associated with it outweigh the disadvantages. Some of the advantages of the

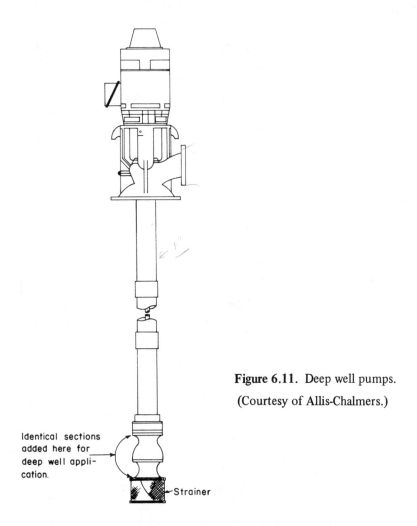

Figure 6.11. Deep well pumps.

(Courtesy of Allis-Chalmers.)

Identical sections added here for deep well application.

Strainer

centrifugal pumps are low capital cost, small space required, low operation cost, minimum of operational trouble, and constant flow of water.

6.4 PUMP CHARACTERISTICS

Contrary to a reciprocating pump, the discharge through a centrifugal pump is dependent on the head to which water is pumped. The centrifugal pump has a maximum efficiency corresponding to a particular value of discharge Q, head H, and speed N. The curves drawn for discharge versus H, N and efficiency e are known as characteristic curves (Figure 6.12). The head delivered by a pump equals the static lift and the head loss due to friction in the suction and delivery pipes. When water is pumped in a piping system, the

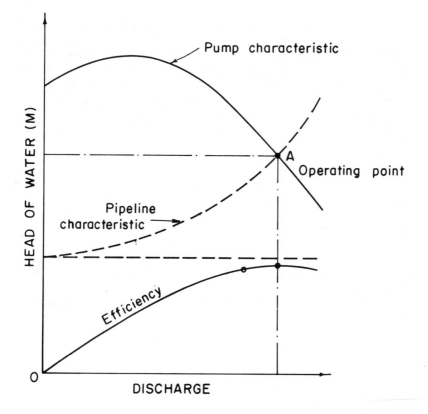

Figure 6.12. Performance characteristics of single centrifugal pump.

geodetic head generally remains constant but the head loss due to friction varies with Q^2. The relationship between the manometric head and the discharge for the pipeline is shown in Figure 6.12. The intersection point A of the two curves indicates the operating head and the corresponding flow rate of the pump. The most suitable pump is one whose efficiency is maximum corresponding to these particular values of head and flow rate obtained by the intersection of the pump characteristics and the pipeline characteristic curves. If this is not the case, then the pump is not suited for this particular condition. Slight shifting of the operating point is possible by changing the geodetic head, without adversely influencing the economics of the piping system.

6.5 OPERATION OF MORE THAN ONE PUMP

To achieve flexibility in operation, more than one pump may be installed. The pumps may operate either in parallel, or in series, or both. Possible arrangements of pumps along with their characteristic curves are shown in Figure 6.13. This figure shows that by suitably arranging the pumps the delivery head as well as the discharge can be increased. When pumps are placed in parallel the discharge increases, and when placed in series the delivery head increases. If the characteristics of the delivery pipe are known, the most suitable arrangement of pumps can be ascertained in a manner similar to that discussed in Section 6.4.

Characteristic curves for a centrifugal pump at various speeds are shown in Figure 6.14. Constant efficiency lines are also drawn in the figure. These curves show that by varying the speed of the impeller, the flow in the pipeline may be controlled.

6.6 AIR LIFT PUMP

An air lift pump may be used in wells with a considerable depth of water. An eductor pipe is inserted in the well so that its lower end is completely submerged (Figure 6.15). Compressed air is forced through a pipe inside the eductor pipe resulting in the formation of an air-water mixture. As

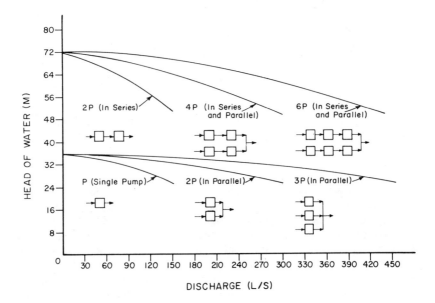

Figure 6.13. Performance characteristics of centrifugal pumps in series and parallel.

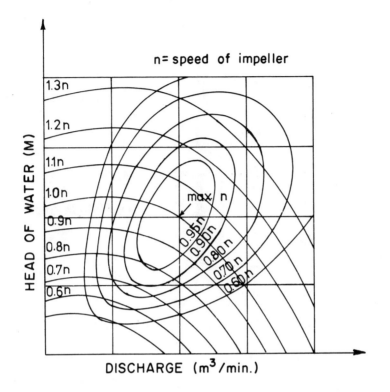

Figure 6.14. Pump characteristics and lines of equal efficiency.

this mixture, lighter than the surrounding water, rises in the pipe, the surrounding water is forced up through the eductor pipe. This causes a continuous flow of water in an upward direction through the eductor pipe.

The height to which water can be raised may be found by considering the work done. Assuming isothermal expansion, the work done by the compressed air diffused into water = $\int p\,dV$ in which p is the absolute pressure and V is the volume of air. But pV = constant = $p_a V_a = p_c V_c$ in which subscripts a and c are used for atmospheric air and compressed air.

Hence
$$p = \frac{p_a V_a}{V}$$

Substituting this in the above expression for work done, we get

$$\text{Work done} = p_a V_a \int_{V_c}^{V_a} \frac{dV}{V} = p_a V_a \ln\left(\frac{V_a}{V_c}\right) = p_a V_a \ln\left(\frac{p_c}{p_a}\right)$$

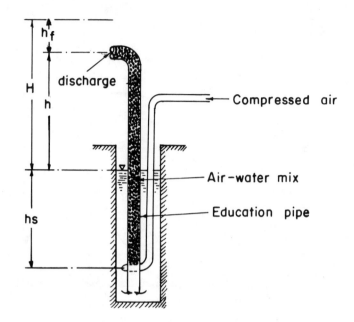

Figure 6.15. Air lift pump. (Courtesy of TH-Dresden.)

Work done/sec $= p_a \dfrac{V_a}{t} \ln \dfrac{p_c}{p_a} = p_a Q_a \ln \dfrac{h_s + h_a}{h_a}$

This power is used to raise the air water mixture $(Q_a + Q_w)$ against a head of $h_t = h_s + h + h_f$ (Figure 6.15).

Thus $p_a Q_a \ln \dfrac{h_s + h_a}{h_a} = (Q_a + Q_w) \rho g h_t$

$$p_a \ln \dfrac{h_s + h_a}{h_a} = \dfrac{Q_a + Q_w}{Q_a} \rho g h_t \quad \cdots \cdots \cdots \cdots (6.7)$$

From Equation 6.7 it is evident that if the height to which water is raised needs to be increased, then either there should be a corresponding increase in the depth the pipe is submerged or the quantity of air supplied. The minimum value of the ratio of H and h_s should be 1:3. The greatest disadvantage of an air lift pump lies in the fact that there has to be sufficient submergence which may make the well unnecessarily deeper.

The efficiency of an air lift pump may be between 15 and 40 percent. It may discharge water between 1 ℓ/s and 120 ℓ/s. The pump has no moving parts and is therefore suitable for water carrying sand and solid particles.

6.7 GUIDE FOR SELECTION OF PUMPS

1.   For flows up to 1,895 ℓ/min (500 gpm) use two pumps (one working and the other as a stand-by).
2.   For flows more than 1,895 ℓ/min (500 gpm) but less than 5,685 ℓ/min (1,500 gpm) use three pumps (two working and the third as a stand-by).
3.   For flows more than 5,685 ℓ/min (1,500 gpm) but less than 11,370 ℓ/min (3,000 gpm) use four pumps (three working and the fourth as a stand-by).
4.   For flows more than 11,370 ℓ/min (3,000 gpm) use six pumps (four working and two as stand-by).

Example 6.2

Information regarding pipelines and pumps is given below:
1.   150 mm diameter pipe has a length of 64 m. Its equivalent length for friction purposes is 107 m. It costs Iraqi Dinar 4.50 per m length.
2.   200 mm diameter pipe has a length of 64 m. Its equivalent length for friction purposes is 92 m. It costs Iraqi Dinar 6.00 per m length.
3.   Pump A costs Iraqi Dinar 700.
4.   Pump B costs Iraqi Dinar 400.
5.   Pipe and pump characteristics are given in Figure 6.16.
Determine which of the following combinations will be suitable and economical for a flow of 62 ℓ/s.
   (a)   Using pump A and 200 mm dia. pipe
   (b)   Using pump A and 150 mm dia. pipe
   (c)   Using pump B and 200 mm dia. pipe
   (d)   Using pump 2 B and 150 mm dia. pipe

Solution:

   (a)   Cost = 700 + 64 x 6 = 700 + 384 = I.D. 1084
   (b)   Cost = 700 + 64 x 4.5 = 700 + 288 = I.D. 988
   (c)   Cost = 400 + 64 x 6 = 400 + 384 = *I.D. 784*—most economical
   (d)   Cost = 2 x 400 + 64 x 4.5 = 800 + 288 = I.D. 1088

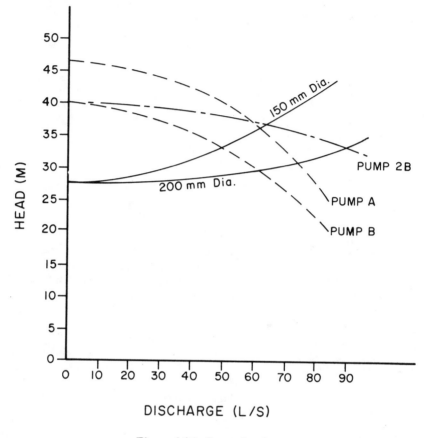

Figure 6.16. Example 6.2.

7. Distribution System

7.1 PATTERNS OF WATER DISTRIBUTION

A network of pipes is used to distribute water to a community. There are various ways in which this can be done, e.g. a branching pattern with dead ends, a grid pattern, and a grid pattern with a large diameter pipe loop around an area with a high demand for water. Often a combination of these systems is used in a city.

7.1.1 Branching Pattern with Dead-Ends

A branching pattern with dead-ends is similar to the branching of a tree (Figure 7.1a). To the trunk line (primary feeders), mains (secondary feeders) are connected and to the mains, submains (small distribution mains) are joined for supplying water to the buildings. In pipes with dead-ends the flow of water is always in the same direction, and water is supplied to an area by a single pipe.

Advantages: a) Branching is a very simple method of water distribution; b) The design of such a pipe network is simple; and c) the required dimensions of the pipes are economical.

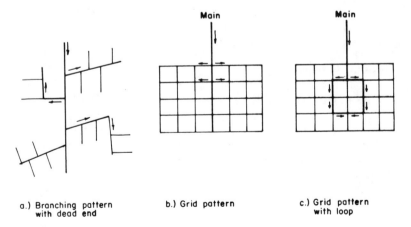

a.) Branching pattern with dead end

b.) Grid pattern

c.) Grid pattern with loop

Figure 7.1. Pattern of water distribution.

97

Disadvantages: a) Sediments accumulate due to stagnation of the dead-end, occasionally causing tastes and odors if the pipe is not regularly flushed; b) the area receiving water from a pipe under repair is without water until the work is completed; and c) insufficient water pressure may occur when additional areas are connected to the water supply system.

7.1.2 Grid Pattern

In a grid pattern all the pipes are interconnected (Figure 7.1b) with no dead-ends. Water can reach any point from more than one direction.

Advantages: a) Water in the supply system is free to flow in more than one direction and stagnation does not occur as readily as in the branching pattern; b) in case of repair or a break in a pipe, the area connected to that pipe will continue to receive water, as water will flow to that area from the other side; and c) there will be little adverse effect on the supply of water due to large variations in water consumption.

Disadvantages: a) The calculation of pipe sizes is more complicated; b) more pipe and fittings are required.

7.1.3 Grid Pattern with Loops

Loops can be provided in a grid pattern (Figure 7.1c) to improve water pressure in portions of a city (industrial, business, and commercial areas). Loops should be strategically located so that as the city develops the water pressure will be sustained.

The advantages and disadvantages of this pattern are the same as those of the grid pattern.

7.2 PRESSURE IN PIPES

In residential areas having houses to four stories high, the pressure in the pipes should be between 1.8×10^5 to 2.8×10^5 N/m^2 (1.8 to 2.8 kg/cm^2). Pipelines supplying water to fight fires must have a pressure of 4.2×10^5 N/m^2 (4.2 kg/cm^2). Pressure of about 5.3×10^5 N/m^2 (5.3 kg/cm^2) is required for business areas.

7.3 VELOCITIES AND SIZE

Velocities in the pipes should not exceed 0.6 to 1.2 m/sec. Pipelines 10 cm in diameter or larger are used to transmit water in a city. In large cities the size of the distribution mains should not be less than 15 cm (6 in.). Pipes of standard sizes are available and must be used for economical reasons.

7.4 GENERAL REQUIREMENTS FOR PIPE WORK

Mains should be divided into sections and valves should be provided in each, so that any section may be taken out of operation for repairs. Dead-ends are to be avoided. If a dead-end is a must, a hydrant should be provided for cleaning. Air valves at summits and drains at low points between summits should be installed.

Mains should follow the general contour of the ground. Pipes should not ordinarily rise above the hydraulic gradient (Figure 7.2), otherwise there may be problems with siphonage. The minimum cover under roadways should be 90 cm (3 ft) and under foot paths 75 cm (2.5 ft).

Proper maintenance and operation of a water supply system requires that a number of appurtenances be provided in the pipeline. Gate valves are generally used at the summit points and to isolate a particular section of pipeline. Sluice gates are generally used in pipelines laid at steep grades or in openings into wells. At low points in a pipeline, blow-off valves or scour valves are used to discharge the accumulated sediment.

At summits air tends to accumulate due to the release of air from water and displacement of air from the pipe. Automatic air release valves are provided to remove accumulated air. Pipes constructed of steel and other flexible materials must have valves which automatically allow air to enter when the pipeline is emptied in order to prevent a vacuum which will cause the pipe to collapse.

7.5 PIPE JOINTS

7.5.1 Bell and Spigot Joint (Figure 7.3a)

A bell and spigot joint is used both for cast iron and steel pipes. The spigot end of the pipe is pushed inside the bell end. A tarred gasket made of cotton yarn, or jute is packed into the open space between the bell and the spigot. A strap is placed around the joint, and molten lead is poured in and tamped. The joint is finished as shown in Figure 7.3a.

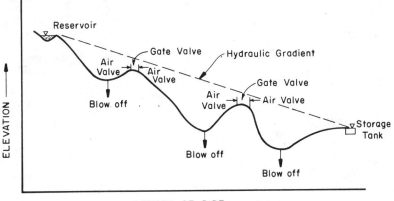

Figure 7.2. Profile of pipeline.

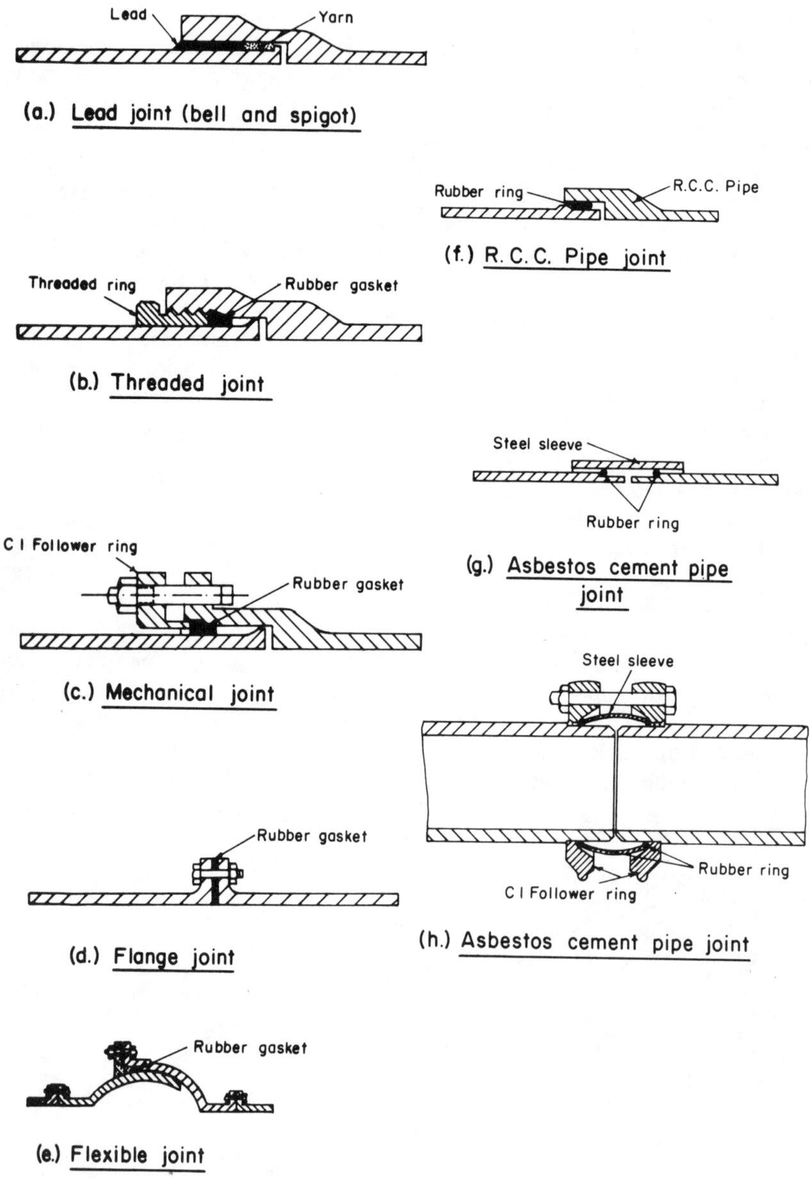

(a.) Lead joint (bell and spigot)

(b.) Threaded joint

(c.) Mechanical joint

(d.) Flange joint

(e.) Flexible joint

(f.) R.C.C. Pipe joint

(g.) Asbestos cement pipe joint

(h.) Asbestos cement pipe joint

Figure 7.3. Pipe joints.

7.5.2 Threaded Joint (Figure 7.3b)

Threaded joints in pipes are strong. A rubber gasket is used to make the joint water tight. The bell end is threaded on the inside to fit with an outside threaded ring. This ring presses against the rubber gasket making the joint water tight.

7.5.3 Mechanical Joint (Figure 7.3c)

A rubber gasket of trapezoidal cross section is pressed against the spigot end. A cast iron follower ring is connected to the bell end making the joint water tight.

7.5.4 Flange Joint (Figure 7.3d)

A flange joint is suitable for pipes under high pressure and for pipes subject to variations in temperature. Various fittings in a pump house are connected by flange joints. Rubber or similar material gaskets 3-5 mm (1/8 - 1/4 in.) thick are placed between flanges which are connected by bolts.

7.5.5 Flexible Joint (Figure 7.3e)

The direction and the slope of pipes connected by a flexible joint can be varied up to a maximum of $20°$. This joint is especially suitable for pipes being placed under water.

7.5.6 Welded Joint

Large diameter steel pipes are generally welded together. Frequently small diameter steel pipes are welded together. Welded joints require greater skill than the ones mentioned above, and careful quality control is required.

7.5.7 Concrete Pipe Joint (Figure 7.3f)

Concrete pipes with bell and spigot joints are generally used. A rubber ring is used to make joints water tight. Such joints are used for water pipes which are not under pressure.

7.5.8 Asbestos Cement Pipe Joint (Figure 7.3g,h)

This joint consists of a steel or cast iron sleeve which fits over the ends of the pipes. It is made water tight with two rubber rings placed between the sleeve and the pipes. Sleeves are frequently constructed of the same material as the pipe.

7.6 VALVES

Valves of various types are used in a water supply system. Details of three commonly used valves are given below.

7.6.1 Gate Valve

Gate valves are used in every main and submain (Figure 7.4) to isolate a portion of the system during repair. Valves are provided in both intersecting pipelines. Generally they are used at intervals of not more than 160 - 250 m (500 - 800 ft). Manholes or chambers are constructed at each valve location to provide easy access.

A large valve has an auxiliary valve or a by-pass to allow the flow of a small stream of water. These are required when there is a considerable difference in pressure on the two sides of the valve. When opening the main valve, the auxiliary valve is opened first to reduce the pressure and facilitate the opening of the main valve; when closing, the main valve is closed first.

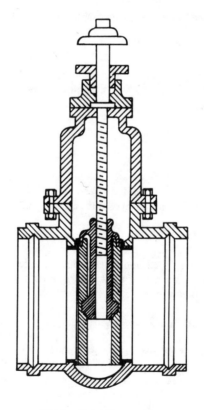

Figure 7.4. Gate valve.

7.6.2 Air Valve

An air valve is used at summits in the pipelines for the removal of air. If an air valve is not provided, the presence of air will reduce the effective area of the pipe. An air valve is shown in Figure 7.5. The float provided in the chamber regulates the opening and the closing of the valve. When the chamber of the valve is full of water the opening remains closed, but when filled with air, the float drops allowing the air to escape through the opening. Consequently water rushes into the chamber and the air space is eliminated.

7.6.3 Check Valve

A check valve allows flow in one direction only. Water flowing in the direction shown in Figure 7.6 causes the flap of the valve to swing on the pivot and open. When flow stops, the flap drops to its seat and prevents flow in the opposite direction.

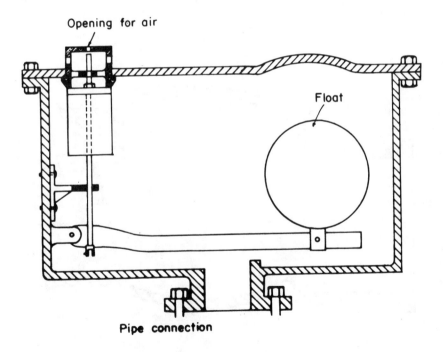

Figure 7.5. Air valve.

7.7 HYDRANT

A hydrant is used on the mains to provide a connection for fire-hoses to fight fire (Figure 7.7). There are two types of hydrants, namely flush and post. The flush hydrant is placed inside a cast iron box which is level with the street surface. The post hydrant projects about one meter above the surface of the street. Hydrants should be placed at easily accessible locations. Valves may be of the gate or compression type and these may be one or more outlets.

7.8 DISTRIBUTION OF WATER IN A BUILDING

7.8.1 Service Pipe

Pipelines are used to supply or remove water from a building. The plumbing system starts at the point where water enters the building from the water main and ends at the point where the drainage water leaves the

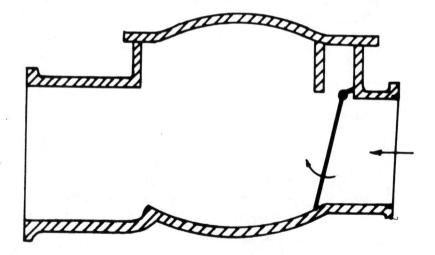

Figure 7.6. Check valve.

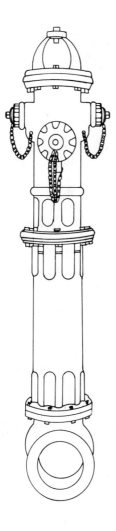

Figure 7.7. Fire hydrants.

building. The pipeline connecting the plumbing system of the building to the water main is called the service pipe. Valves and meters are fitted in the service pipe as shown in Figure 7.8.

7.8.2 Distribution Pipes

Galvanized iron pipes are generally used to supply water in buildings. The building distribution piping system must be designed so that water reaches the upper floors when a number of fixtures are open on lower floors.

If the water pressure is inadequate, a vacuum may be created in the distribution pipes and allow contaminated water to be drawn into the building distribution system from fixtures which do not have flow prevention devices (Figure 7.9).

7.8.3 Design Considerations

It is essential that a building water distribution system be designed so that adequate water is available to all the fittings and fixtures. Water consumption of 135 liters per capita per day may be used in estimating domestic needs. The quantity of water used depends upon the class of building and the standard of living of the inhabitants.

To find the size of pipes and fittings needed for a water supply system, the rate of flow must be known. The rate of flow depends upon the number of hours the water is supplied at high pressure. Where the supply is limited to a few hours in a day, the rate of flow will be large due to simultaneous use of a large number of fittings. The head loss due to friction must be considered when calculating the size of the pipe.

Cross-connections between pipes carrying potable water and water which may contaminate the water supply must not be allowed. Backflow can be prevented in water supply systems by proper sizing of the distributing system. Traps and air seals between the discharge point and the highest level of wastewater possible in the fixture must be provided. Check valves are not to be used to prevent backflow in the system. Pipe work should be completely water tight to prevent waste as well as entry of any substance which may make the water unsafe. Pipes must not pass into or through a sewer, drain, or manhole or any place which may contaminate the water supply. Change in diameter should be gradual so as to minimize head loss.

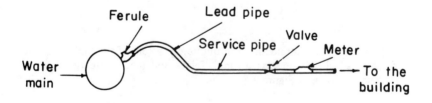

Figure 7.8. Connection of plumbing system with water main.

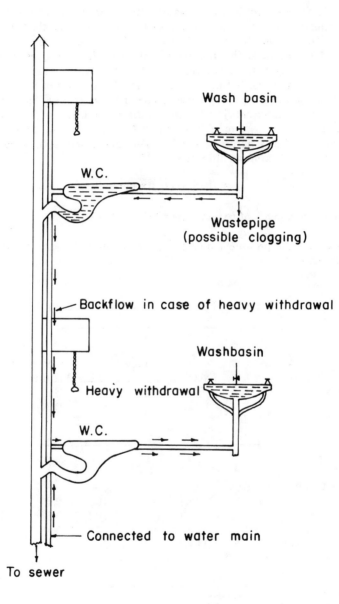

Figure 7.9. Backflow in buildings.

7.8.4 Design of Water Supply Pipes

It is necessary to design the water distribution system in a building so that adequate water at the desired pressure is available in all the fittings and fixtures. To accomplish this the demand on the system and the head losses have to be calculated. The demand can be determined by considering the number and types of fixtures used. Generally water will not be drawn at the same time from all the fixtures in the system. If n fixtures of the same size are installed in the system, the demand should be taken as $\sqrt{n}$ times the rate of flow from each fixture. The demand, or the probable rate of flow, is based on the fixture unit. A rate of flow of 0.25 ℓ/sec from a 3/8 inch (9.5 mm) faucet is defined as a fixture unit (DIN). The demand load is the square root of the number of fixtures multiplied by 0.25, or

$$q = 0.25 \sqrt{\text{no. fixture units}}$$

Therefore, the number of fixture units = $(q/0.25)^2 = 16\,q^2$ . . . . . (7.1)

The flow and the pressure required for the common domestic fixtures are given in Table 7.1.

Table 7.1. Flow and Pressure Requirements[1]

Fixture	Excellent Flow ℓ/min	Flow ℓ/min ISI Code	DIN	Fixture Unit (DIN)	Pressure[a] at Outlet (Faucet Wide Open) kg/cm^2
Lavatory faucet single	15	-	7.5	0.25	0.28
Bathtub faucet single	25.7	4.5	-	-	0.35
Combination bathtub faucet	27	-	-	-	0.35
Sink faucet	25.7	4.5	10.5	0.5	0.35
Shower head	25.7	4.5	10.5	0.5	0.21
Water closet, tank type	19	9	7.5	0.25	0.35
Water closet with flush valve	114	-	36	6	1.75
Garden hose and nozzle	25.7	-	-	-	2.0 at hydrant

DIN — German Standard
ISI — Indian Standard Institution (1 Fixture Unit = 9 ℓ/min = 0.15 ℓ/s)
U.S. Standard — 1 Fixture Unit = 1 cfm = 7.5 gallon per min = 28.5 ℓ/min
[a]Pressure in N/m^2 = kg/cm^2 x 10^5

Example 7.1

Calculate the total fixture units and the demand for the following fixtures in a building: 1 water closet (tank type), 1 wash basin, 1 sink, and 1 shower head.

Total fixture unit = $(0.25 + 0.25 + 0.5 + 0.5) = 1.5$
Demand = $\therefore 1.5 \times 0.25 = 0.31$ ℓ/sec

7.8.5 Distribution System for Buildings

7.8.5.1 Types of Distribution (Figure 7-10)

1. The direct feed type is used when the pressure in the municipal main is adequate to supply all fixtures inside the building with water for 24 hours each day.
2. Indirect feed is used when the pressure in the municipal water main is insufficient to supply water to the fixtures at all times of the day. Water is supplied to the fixtures inside a building by an elevated tank placed on the roof of the building or by a pressurized tank.

7.8.5.2 Methods of Design

1. The same procedure is used in the design of the piping system for a direct and indirect distribution system.
2. Indirect feed.

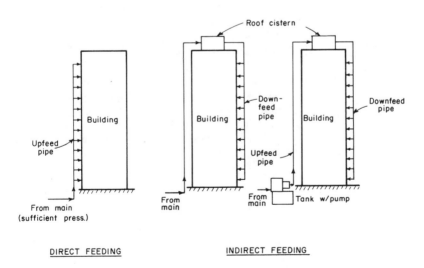

Figure 7.10. Distribution system for buildings.

a.   An up-feed system can be employed under two conditions:
    i.   When the water pressure in the municipal main is adequate to supply the elevated tank with water during part of the day.
        Design criteria
        1.   Minimum duration of pressure to supply the elevated tank should usually be 6 hours in a day.
        2.   Flow in the pipe during the period of adequate pressure must equal the water consumption per day.
        3.   The diameter of the up-feed pipe must be large enough to insure a pressure of not less than 0.68 kg/cm^2 (10 psi) at the elevated tank.
    ii.   When the pressure in the distribution system is not adequate to deliver water to the elevated tank at any time.
        Design criteria
        1.   Use a storage tank at the ground level.
        2.   Use pumps to raise the water from the storage tank to the elevated tank placed on the roof of the building.
        3.   Design the pipe from the distribution system to the ground storage tank to insure an adequate supply of water.
        4.   Select storage tanks with a capacity compatible with the pumping schedule and the variation in the municipal supply.
        5.   Select a suitable pump depending upon the duration of pumping (1 to 3 to 6 hours).
        6.   Design the up-feed pipe from the ground level storage tank to insure a pressure of 0.68 kg/cm^2 at the discharge into the storage tank.
        7.   The volume of the elevated tank should equal that of the ground level storage tank.
b.   Down-feed design.
    i.   The pressure at the fixtures should be near 0.68 kg/cm^2 (10 psi) but not less than 0.28 kg/cm^2 (4 psi).
    ii.   No pipe should be less than 0.5 inch.
    iii. All pipes should be designed to carry less than twice the average rate of flow.

Example 7.2

Design   up-feed   pipe for a three storied building for the following conditions:
No. of flats in each story = one.
Height of each story = 4 m (12 ft).
Average water consumption = 190 ℓ/c/d (app. 50 gpcd).
Equivalent length of pipe in the building = 100 m (300 ft).
Pressure in the distribution system = 0.68 to 2 kg/cm^2 (10 - 30 psi).

Pressure in the municipal system reaches 2 kg/cm^2 for a 4-hour period each day.

Solution:

Total water consumption = (3 stories) (5 people/flat) (190 ℓ/c/d) (1 flat/ story) = 2,850 ℓ/day = 2.85 m^3/day (750 g/d).

Rate of water withdrawl from municipal system = (2,850/4(60)) = 11.88 ℓ/min (3 gpm).

Assume pipe size = 2.54 cm (1 in), hence head loss = 2.5 m/100 m (2.5 ft/100 ft) (Figure 7-11).

∴ For 100 m length, head loss = 2.5 m (8.2 ft) which is equivalent to a pressure of 2.5/10.33 = 0.242 kg/cm^2 (3.2 psi).

Pressure of water at the top of the building = 2 - 0.242 = 1.758 kg/cm^2 (30 - 3.2 = 26.8 psi).

Static pressure of 1.758 kg/cm^2 is more than the static head of 1.162 kg/cm^2. Therefore the 2.54 cm (1 in) diameter pipe is adequate.

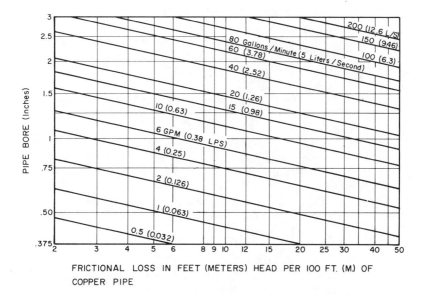

FRICTIONAL LOSS IN FEET (METERS) HEAD PER 100 FT. (M.) OF COPPER PIPE

Figure 7.11.   Pipe sizing graph.

Example 7.3

Design the distribution system in a building assuming that the pressure in the city main is not adequate to deliver water to a storage tank on the roof of the building. The following conditions exist:

Number of persons in the building = 100.
Water consumption = 280 ℓpcd (74 gpcd).
Assume duration of pumping = 1 hour.
Equivalent length of pipe from the main to the ground level storage tank = 141 m (462 ft).
Total height of the building (three floors) = 15 m (49.2 ft).
Minimum water pressure in municipal main = 2 kg/cm^2.

Solution:

1.   Ground level storage tank Volume = 280 x 100 = 28 m^3 (7,400 gallons = 989.38 ft^3).
     Assume the depth = 3 m (9.84 ft).
     Area = 28/3 = 9.4 m^2 (98.94 ft^2).
     Tank dimension = 3.1 m x 3.1 m x 3 m (9.94 ft x 9.94 ft x 9.84 ft).
     Rate of flow = (280 x 100)/(24 x 60) = 19.4 ℓ/m (5.12 gpm).
     Assume 2.54 cm (1 in) diameter pipe supplies water from the main.
     Loss of head = 6 m/100 m (6 ft/100 ft).
     Total loss of heat = 6 x 1.41 = 8.46 m (28.2 ft) which is equivalent to a pressure of 8.46/10 = 0.85 kg/cm^2 (12.22 psi).
     Depth of water in the tank = 3 m (9.84 ft) = 0.3 kg/cm^2 (4.3 psi).
     ∴ Head against which main has to supply water to the tank = 0.85 + 0.3 = 1.15 kg/cm^2 (16.55 psi) which is less than 2 kg/cm^2.
2.   Roof cistern (elevated tank).
     Size = size of the ground tank = 3.1 m x 3.1 m x 3 m (9.94 x 9.94 x 9).
3.   Pumps:
     Use two pumps, one working and one stand-by.
     Capacity of pump (one hour of operation) = 28,000/60 = 466.7 ℓ/min (123.1 gpm).
     Assume the diameter of up-feed pipe = 7.6 cm (3 in).
     Head loss = 10 m/100 m (10 ft/100 ft).
     Static head = 15 + 3 = 18 m (49.2 + 9.84 = 59.04 ft).
     Equivalent length for bends, valves, etc. = 61 m (200 ft).
     ∴ Total equivalent length = 18 + 61 = 79 m (59.04 + 200 = 259.04 ft).

∴ Total head loss due to friction = 7.90 m (25.9 ft).
Residual pressure at the end of pipe = 6.8 m (22.3 ft).
∴ Head against which the pump should work = 7.90 + 6.8 = 14.7 m (25.9 + 22.3 = 48.2 ft).
Power required = DgQH = 1,000 x 9.81 x (0.467/60) x 14.7 = 1.12 watts or 1.53 m hp

4.   Down-feed pipe:
Average water consumption = 19.4 ℓ/min (5.1 gpm).
Maximum water consumption = 38.8 ℓ/min (10.2 gpm).
Use 4.0 cm (1.5 in) diameter pipe.
Head loss = 2.7 m/100 m (2.7 ft/100 ft).
Assume equivalent length of pipe, bends, valves, etc. = 10 m.
∴ Total loss of head = 0.27 m (0.89 ft).
Head at which water is supplied to the top floor = (height of the story + height of water in the tank) - head loss = (5 + 3) - 0.27 = 7.73 m (25.31 ft) = 0.773 kg/cm² (10.93 psi) which is satisfactory.

The supply pressure on the first and second floor will be higher than 0.773 kg/cm² because the difference in elevation between floors will exceed head losses in the supply pipe.

7.9 MATERIALS FOR PIPES

The following materials may be used for pipes:
Cast iron
Steel coated with bitumen or any composition of bitumen
Galvanized iron
Reinforced concrete
Copper
Brass
Asbestos cement
Polyethylene and polyvinyl chloride (PVC)

7.10 SELECTION OF PIPE

Some of the important factors which should be considered in the selection of pipe material are given below: a) Durability, type of liquid to be transported; b) Resistance to corrosion and erosion, strength; c) Cost of pipe, cost of handling and lay out; d) Type of joint, its water tightness and assembly; and e) Site conditions, availability, local material, and cost of maintenance.

8. Design of Pipelines

8.1 HYDRAULICS OF FLOW IN PIPES

The head loss in a pipeline carrying water can be calculated by the well known Darcy-Weisbach equation

$$h_f = \frac{f\ell v^2}{2gd} \quad \ldots \ldots \ldots \ldots \ldots \ldots \ldots \quad (8.1)$$

in which

h_f	=	head loss
f	=	resistance coefficient
ℓ	=	length of pipe
d	=	diameter of pipe
v	=	velocity
g	=	acceleration due to gravity

Darcy's equation can be transformed to obtain Chezy's equation as follows:

$$v^2 = \frac{2gd}{f\ell} h_f$$

$\dfrac{h_f}{\ell}$ = slope of the energy line or the hydraulic gradient = S

For a circular pipe flowing full hydraulic mean radius = R = A/P = d/4.

A = cross sectional area of pipe = $\pi d^2/4$

P = wetted perimeter of pipe = πd

$$\therefore v^2 = \frac{8g}{f} RS$$

or

$$v^2 = C^2 RS \qquad \text{in which} \quad C^2 = \frac{8g}{f}$$

Therefore, $V = C\sqrt{RS}$ $\ldots \ldots \ldots \ldots \ldots \ldots$ (8.2)

In Chezy's equation the value of C must be known. Manning and Strickler developed the following equation from the Chezy equation. The Manning-Strickler equation is used most frequently in practice.

115

Manning-Strickler equation: $V = \frac{1}{n} R^{1/6} \cdot R^{1/2} \cdot S^{1/2}$ (in which $C = \frac{1}{n} R^{1/6}$)

$$= \frac{1}{n} R^{2/3} S^{1/2} \qquad \ldots \ldots \ldots \quad (8.3)$$

in which
    n   =   roughness coefficient

For design calculations nomograms may be used such as the one given in Figure 8.1.

    If the value of f in Equation 8.1 is known, the value of Chezy's constant C can be determined. The Prandtl-V. Karman-Colebrook equations (Equations 8.4, 8.5, 8.6) can be used to find f.

Hydraulically smooth zone:

$$\frac{1}{\sqrt{f}} = 2 \log \left(R_e \frac{vf}{2.5\ell} \right) \qquad \ldots \ldots \ldots \quad (8.4)$$

Transition zone:

$$\frac{1}{\sqrt{f}} = -2 \log \left(\frac{2.5\,\ell}{R_e \sqrt{f}} + \frac{k}{3.71d} \right) \qquad \ldots \ldots \quad (8.5)$$

Hydraulically rough zone:

$$\frac{1}{\sqrt{f}} = 2 \log \left(\frac{3.71d}{k} \right) \qquad \ldots \ldots \ldots \quad (8.6)$$

in which
    f =   friction factor
    k =   absolute roughness (m)
    d =   diameter (m)
    k/d =   relative roughness
    R_e =   Reynolds number = Vd/v
in which
    V =   velocity of flow in pipe (m/sec)
    v =   kinematic viscosity of water = 1.206×10^{-2} (cm²/sec)
               [1.206×10^{-6} (m²/sec)] at 13°C

Values of f can also be determined from Figure 8.2 which is based on Equations 8.4, 8.5, and 8.6.

    When using the Prandtl-Karman equations, proper estimation of the k value is difficult. The value of k may be determined experimentally. It varies considerably for various types of surfaces. Values of k for various pipe materials are given in Table 8.1.

    It should be noted that k-values can be considerably higher for old pipes. Results of investigation on pipes under operating conditions are given in Table 8.2. The head losses due to pipe appurtenances, bends, valves, etc., are included in the following tables of k values. The Manning equation is more popular than the Prandtl equation because of the difficulty in determining the value of f.

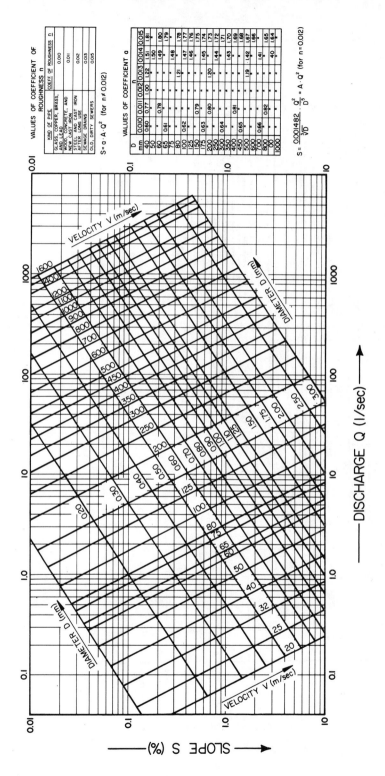

Figure 8.1. Head losses in pipes by Manning's formula for n = 0.012.

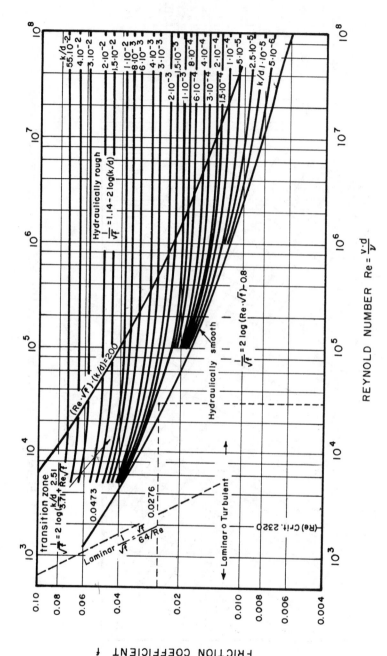

Figure 8.2. Moody diagram for friction in pipes.

Table 8.1. Value of k (mm) for New Pipes[1]

Brass	0.0015	Galvanized iron	0.15
Copper	0.0015	Steel (common & welded)	0.1
Concrete	1.2	Riveted steel	1.8
Cast iron:		Wood stave	0.6
– Uncoated	0.25	Steel and cast iron pipes	
– Asphalt dipped	0.125	with incrustations	1.6
– Cement lined	0.0024	Asbestos cement pipe	0.1
– Bituminous lined	0.0024		
– Centrifugally spun	0.003		

Table 8.2. Values of k (mm) for Old Pipe[1]

Type of Pipes	Raw Water, k (mm)	Treated Water	
		Mains k (mm)	Sub-main k (mm)
Steel and cast iron pipe	max. - 6.0	2.0	3.0
Asbestos cement pipe	0.4 - 1.5	0.4	0.4
Spun hume pipe	1.0 - 1.4	1.0	1.0

In Equations 8.4 and 8.5, f is on both the right and left hand sides of the equations. This makes the solution difficult. Therefore tables, diagrams (Figures 8.3 and 8.4), and simpler methods have been developed.

Liebhold[2] has suggested a sufficiently accurate equation which is a simplified form of Prandtl's equation.

$$\frac{1}{\sqrt{f}} = C_1 + 2 \log d \ . \ . \ . \ . \ . \ . \ . \ . \ . \ . \ . \ . \ \ (8.7)$$

in which

$$C_1 = 9.14 - 2 \log \frac{10^4 \ kV + 1}{V} \ . \ . \ . \ . \ . \ . \ . \ . \ \ (8.8)$$

The value of the constant C_1 has been found for water temperatures ranging from 5°C to 15°C and an average value of $1/\sqrt{f}$ is 8.2 (Table 8.3).   In tropical countries the water temperature ranges from 10° to 25°C. For this range of

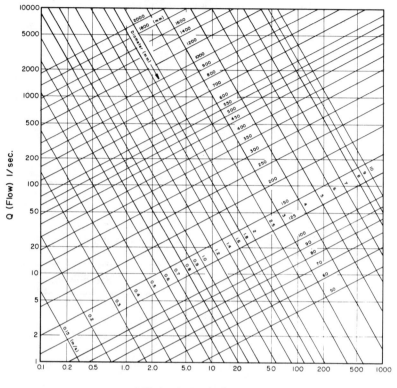

S (Hydraulic gradient) m per 1000 m

Figure 8.3. Nomogram for flow of water in pipes (K = 0.1 mm in Prandtl-Karman-Colebrook equation).

temperatures the average value of $1/\sqrt{f}$ is 8.15. As there is little difference in the value of $1/\sqrt{f}$, Equation 8.7 may also be used for 10° to 25°C.

The range of values for k usually encountered in water supply is   0.1 to 0.4 mm, and for sewerage k is approximately 1.5 mm. After finding the value of f, Chezy's constant can be calculated as follows:

$$C = \sqrt{\frac{8g}{f}}$$

Example 8.1: Calculate the flow in pipes shown in Figure 8.6

Let   k = 0.4 mm
      d = 100 mm $= \dfrac{1}{10}$ m
      V = 1.0 m/s

Example 8.1: Calculate the flow in pipes shown in Figure 8.6

Circuit No.	Pipe	Length m	Dia. mm	Q_o ℓ/s	h_{f_o}/ℓ %	h_{f_o} cm	$\dfrac{h_{f_o}}{Q_o}$	ΔQ_o	Q_1 ℓ/s	Head Loss Between A and D
1	AB	2000	250	+30	0.21	+420	0.014	-0.06	+29.94	via AB, BC,
	BC	1000	150	+10	0.37	+370	0.037	-0.06	+ 9.94	CD = 1.6 m
	CF	2000	150	- 5	0.094	-188	0.0376	-0.06 -0.051	- 5.11	via AF, FE,
	FA	1000	250	-50	0.59	-590	0.0118	-0.06	-50.06	ED = 1.61 m
					Sum	+ 12	0.1004			

$$\Delta Q = - \frac{12}{2 \times 0.1004} = -60 \ \text{cm}^3/\text{s} = -0.06 \ \ell/\text{s}$$

Circuit No.	Pipe	Length m	Dia. mm	Q_o ℓ/s	h_{f_o}/ℓ %	h_{f_o} cm	$\dfrac{h_{f_o}}{Q_o}$	ΔQ_o	Q_1 ℓ/s
2	FC	2000	150	+ 5	0.094	+188	0.0376	+0.051 +0.06	+ 5.11
	CD	1000	150	+15	0.81	+810	0.054	+0.051	+15.051
	DE	2000	125	- 5	0.26	-520	0.104	+0.051	- 4.949
	EF	1000	200	-25	0.5	-500	0.02	+0.051	-24.949
					Sum	- 22	0.2156		

$$\Delta Q = - \frac{-22}{2 \times 0.2156} = +51 \ \text{cm}^3/\text{s} = +0.051 \ \ell/\text{s}$$

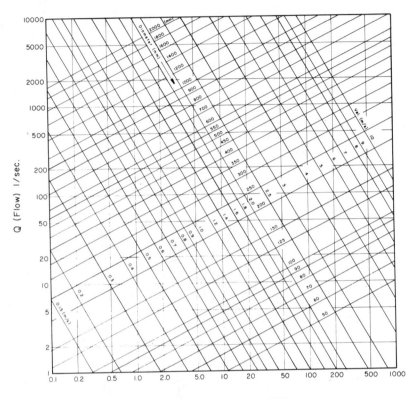

Figure 8.4. Nomogram for flow of water in pipes (K = 0.4 mm in Prandtl-Karman-Colebrook equation).

From Table 8.3

$$C_1 = 7.74$$

$$\frac{1}{\sqrt{f}} = 7.74 + 2 \log d$$

$$= 7.74 + 2 \log \frac{1}{10} = 5.74$$

$$f = \sqrt{\frac{1}{5.74}} = 0.4174$$

$$C = \sqrt{\frac{8g}{f}} = \sqrt{\frac{8 \times 9.81}{0.4174}} = 13.7$$

Table 8.3. Values of C_1, for Simplified Calculation of Head Loss by Prandtl-V. Karman-Colebrook Equation[3]

Absolute Roughness 10^4 k (m)	Velocity V (m/s)									
	0.2	0.4	0.6	0.8	1.0	1.2	1.5	2.0	2.5	3.0
0.0	7.74	8.34	8.69	8.94	9.14	9.30	9.49	9.74	9.93	10.09
0.5	7.66	8.18	8.47	8.65	8.80	8.89	9.01	9.14	9.22	9.30
1.0	7.58	8.05	8.29	8.44	8.54	8.61	8.70	8.79	8.85	8.89
1.5	7.51	7.94	8.14	8.26	8.34	8.40	8.47	8.54	8.58	8.61
2	7.45	7.83	8.01	8.12	8.19	8.23	8.29	8.34	8.38	8.40
3	7.33	7.66	7.80	7.88	7.94	7.98	8.01	8.05	8.08	8.09
4	7.23	7.51	7.63	7.76	7.74	7.77	7.80	7.83	7.85	7.87
5	7.14	7.39	7.49	7.55	7.58	7.61	7.63	7.66	7.67	7.69
6	7.06	7.28	7.37	7.42	7.45	7.47	7.49	7.51	7.53	7.54
8	6.91	7.10	7.17	7.21	7.23	7.25	7.26	7.28	7.29	7.30
10	6.79	6.95	7.00	7.04	7.06	7.07	7.08	7.10	7.11	7.11
12	6.68	6.82	6.87	6.90	6.91	6.92	6.93	6.95	6.95	6.96
15	6.54	6.65	6.70	6.72	6.73	6.74	6.75	6.76	6.77	6.77
20	6.34	6.44	6.47	6.49	6.50	6.50	6.51	6.51	6.52	6.52
25	6.19	6.26	6.29	6.30	6.31	6.31	6.32	6.33	6.33	6.33
30	6.05	6.12	6.14	6.15	6.16	6.16	6.17	6.17	6.17	6.18

8.2 DESIGN OF DISTRIBUTION SYSTEM

A pipeline system is designed to carry a certain flow rate. The size of pipes must not be unnecessarily over-dimensioned and at the same time there should be sufficient pressure throughout the system.

Basically there are two patterns of pipelines (Section 7.1). One is the branching pattern with dead ends and the other is the grid pattern. In large communities both patterns are generally found. Smaller communities usually have a branching pattern system with dead ends.

8.2.1 Network Analysis (Hardy Cross Method)

Consider a pipe network ABCD in which a quantity of water Q is entering at A and leaving at C (Figure 8.5).   Such a closed network must satisfy the following requirements:

1. At every junction the total quantity of water entering is equal to the algebraic sum of water leaving.

2. In every loop the algebraic sum of head losses through any chosen path is zero.

Either of the following methods can be used to solve a network:

The head may be balanced by correcting assumed flows.

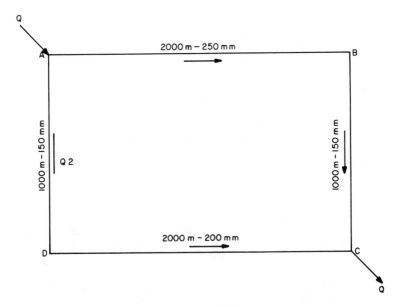

Figure 8.5. Piping system used in Hardy-Cross method.

Or

The flows may be balanced by correcting assumed heads.

The first method is used most frequently. In a network, length, diameter and roughness of the pipes are usually known (Figure 8.6).

In the beginning, we assume that the flow in two sections is Q_1 (+) and Q_2 (-). If the assumption is correct, then the system is hydraulically balanced and we have

$$h_{f_1} - h_{f_2} = 0$$

In general it is not possible to assume such a flow in the beginning. Let us consider that Q_1 and Q_2 have been incorrectly assumed by an amount ΔQ. The corrected flows are $Q_1 + \Delta Q$ and $Q_2 - \Delta Q$. The algebraic sum of the corresponding head losses $h'f_1$ and $h'f_2$ will be zero.

$$h'f_1 - h'f_2 = (K_1 (Q_1 + \Delta Q)^{n'} - K_2 (Q_2 - \Delta Q)^{n'} = 0$$

Expanding we get:

$$(K_1 Q_1^{n'} + k_1 n'Q_1^{n'-1} \Delta Q + ...) - (K_2 Q_2^{n'} - K_2 n'Q_2^{n'-1} \Delta Q + ...) = 0$$

As ΔQ is very small, the terms having powers of ΔQ are neglected. Therefore

$$K_1 Q_1^{n'} + K_1 n'Q_1^{n'-1} \Delta Q - K_2 Q_2^{n'} + K_2 n'Q_2^{n'-1} \Delta Q = 0$$

But

$$hf_1 = K_1 Q_1 n' \quad \text{and} \quad hf_2 = K_2 Q_2 n'$$

i.e. $\dfrac{hf_1}{Q_1} = K_1^{n'-1} \quad \text{and} \quad \dfrac{hf_2}{Q_2} = K_2 Q_2^{n'-1}$

Substituting we have:

$$hf_1 + n' \Delta Q \frac{hf_1}{Q_1} - hf_2 + n' \Delta Q \frac{hf_2}{Q_2} = 0$$

$$n'\Delta Q \left(\frac{hf_1}{Q_2} + \frac{hf_2}{Q_2} \right) = hf_2 - hf_1$$

$$\Delta Q = \frac{hf_1 - hf_2}{n' \dfrac{hf_1}{Q_1} + \dfrac{hf_2}{Q_2}} = \frac{K_1 Q_1 n' - K_2 Q_2 n'}{n' (K_1 Q_1^{n'-1} + K_2 Q_2^{n'-1})}$$

Therefore the general equation for the flow correction for a number of loops is:

$$\Delta Q = \frac{\Sigma hf}{n'\Sigma \dfrac{hf}{Q}} = - \frac{\Sigma K_1 Q^{n'}}{n'\Sigma KQ^{n'-1}} \qquad . . . . \quad (8.9)$$

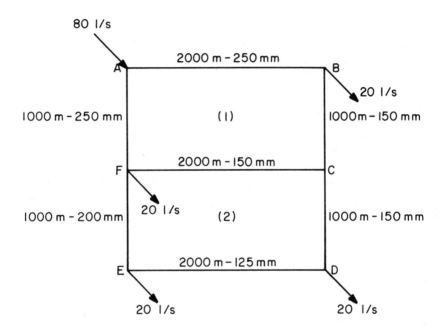

Figure 8.6. Piping system showing characteristics of network.

When using this equation, care should be taken regarding the positive or negative sign of Q_1, Q_2, hf_1, and hf_2. When calculating ΔQ the algebraic sum of the numerator should be used, but the denominator should be added arithmetically. Such a network may not be solved by one approximation, rather a number of successive approximations have to be made until the error in the assumption becomes negligible.

Using the Manning-Strickler equation, the value of h_f will be

$$h_f = \frac{n^2 \, 4^{10/3}}{\pi^2} \, \frac{Q^2}{d^{16/3}} \, L = KQ^2$$

For $n = 0.012$ $\qquad h_f = 0.001482 \frac{1}{d^{5.333}} \, Q^2$

h_f is in m
    L in m
    d in m
    Q in m^3/sec.

When using the Darcy-Weisbach equation we have:

$$h_f = \frac{8f}{n^2 \, gd^5} \cdot \ell \, Q^2 = KQ^2$$

in which

$$K = \frac{8f}{n^2 \, gd^5} \, \ell = K'\ell$$

$$K' = \frac{8f}{n^2 \, gd^5}$$

The following steps will be useful when solving a network.
1. A proper table should be prepared which will facilitate the calculation (Example 8.2).
2. Fill the columns with values given for the network.
3. Assume the flow in the respective pipelines.
4. Determine the first correction ΔQ.
5. The losses probably will not balance after the first correction. Repeat the procedure, arriving at the second correction. Continue repeating the process till the correction becomes negligible.

8.2.2. Design of a Dead End System

Water is supplied from a reservoir at an elevation of 200 m, (Figure 8.8). The elevations of various points in the pipeline are given in brackets. Design the pipelines RA, AA_2, and BC. Assume the minimum pressure in pipes in residential areas must be 35 m of water and in business districts 50 m of water.

Pressure in pipe + elevation at A = 35 + 100 = 135 m
Pressure in pipe + elevation at B = 35 + 70 = 105 m
Pressure in pipe + elevation at C = 50 + 55 = 105 m
Pressure in pipe + elevation at D = 35 + 54 = 89 m

Design of Pipe RA

Total available head = 200 - 135 = 65 m.
Allowable loss of head per 100 m = 100 x (65/3200) = 2.03 m.

Example 8.2. Calculate the discharge in the pipe system shown in Figure 8.7.

Circuit Number	Pipe Line	Length m	Dia. mm	First Trial Q_0 ℓ/sec	h_f/L %	b_{f_0} cm	h_{f_0}/Q_0	ΔQ_0	Second Trial Q_1 ℓ/sec	h_f/L %	b_{f_1} cm	h_{f_1}/Q_1	ΔQ_1	Third Trial Q_2 ℓ/sec	h_f/L %	b_{f_2} cm	h_{f_2}/Q_2	ΔQ_2	Fourth Trial Q_3 ℓ/sec	h_f/L %	b_{f_3} cm	h_{f_3}/Q_3	ΔQ_3	Q_4 ℓ/sec	Head Loss Between A and E
1	AB	1000	200	+15	0.180	+180.0	0.0120	+0.693	+15.693	0.190	+190.0	0.01210	+0.37	+16.063	0.2	+200	0.0125	+0.308	+16.371	0.21	+210	0.01285	0	+16.371	AB + BC + CD
	BK	1300	200	+6	0.029	+37.7	0.0063	+0.693	+5.253	0.025	+32.5	0.00620	+0.37	+4.925	0.021	+27.3	0.0053	+0.308	+5.037	0.022	+27.6	0.0055	0 / −0.141	+4.996	+ DE = 428.3 cm
	KJ	1000	200	−10	0.080	−80.0	0.0080	+0.693	−8.972	0.065	−65.0	0.00725	+0.37	−9.200	0.068	−68	0.0074	+0.308	−9.146	0.066	−66.0	0.0072	0 / −0.171	−9.317	
	JA	1300	250	−25	0.142	−184.6	0.0074	+0.693	−24.307	0.140	−182.0	0.00750	+0.37	−23.937	0.138	−179.4	0.0075	+0.308	−23.632	0.13	−169	0.00715	0	−23.632	AJ + JG + GF
				Σ −46.9			0.0337	$\Delta Q_0=\dfrac{-46.9}{2\times0.0337}=+693\text{ cm}^3/\text{sec}=+0.693$ ℓ/sec	Σ −24.5			0.03305	$\Delta Q_1=\dfrac{-24.5}{2\times0.03305}=+370\text{ cm}^3/\text{sec}=+0.37$ ℓ/sec	Σ −20.1			0.0327	$\Delta Q_2=\dfrac{-20.1}{2\times0.0327}=+308\text{ cm}^3/\text{sec}=+0.308$ ℓ/sec	Σ 0			0.02705	$\Delta Q_3=0$		+ FE = 489.5 cm
4	JK	1000	200	+10	0.080	+80	0.0080	−0.335	+8.972	0.065	+65	0.00725	+0.6	+9.200	0.068	+68	0.0074	+0.254	+9.146	0.067	+67	0.00732	+0.171	+9.317	
	KF	1300	200	+7	0.180	+234	0.0334	−0.335	+5.755	0.120	+156	0.02750	+0.6	+5.993	0.139	+170.7	0.0285	+0.254	+5.895	0.1275	+165.75	0.0282	+0.171	+5.915	
	FG	1000	200	−7	0.038	−38	0.0054	−0.335	−7.416	0.043	−43	0.00580	+0.6	−6.816	0.037	−37	0.0054	+0.254	−6.562	0.0357	−35.7	0.00542	+0.171 / −0.151	−6.391	
	GJ	1300	200	−15	0.180	−234	0.0156	−0.335	−15.416	0.192	−247	0.05655	+0.6	−14.616	0.179	−232.7	0.0157	+0.254	−14.562	0.166	−215.8	0.0149	+0.171	−14.391	
				Σ +42			0.0624	$\Delta Q_0=\dfrac{+42}{2\times0.0624}=-335\text{ cm}^3/\text{sec}=-0.335$ ℓ/sec	Σ −69			0.0566	$\Delta Q_1=\dfrac{-69}{2\times0.0566}=+600\text{ cm}^3/\text{sec}=+0.6$ ℓ/sec	Σ −31			0.057	$\Delta Q_2=\dfrac{-31}{2\times0.057}=+254\text{ cm}^3/\text{sec}=+0.254$ ℓ/sec	Σ −19			0.05584	$\Delta Q_3=\dfrac{-19}{2\times0.05584}=+171\text{ cm}^3/\text{sec}=+0.171$ ℓ/sec		
2	BC	1000	200	+9	0.065	+65	0.0072	+1.44	+10.44	0.084	+84	0.008	+0.668	+11.108	0.095	+95	0.00855	+0.226	+11.334	0.1	+100	0.00884	+0.141	+11.475	
	CD	1300	200	+5	0.02	+26	0.005	+1.44	+6.44	0.033	+42.9	0.00665	+0.668	+7.108	0.041	+52	0.00742	+0.226	+7.334	0.044	+57.2	0.00782	+0.141	+7.475	
	DK	1000	200	−6	0.18	−180	0.0257	+1.44	−6.47	0.158	−158	0.0245	+0.668	−6.294	0.141	−143	0.023	+0.226	−6.290	0.143	−143	0.0228	+0.141	−6.30	
	KB	1300	200	−6	0.029	−37.7	0.0063	+1.44	−5.253	0.022	−28.6	0.00543	+0.668	−4.955	0.02	−26	0.00525	+0.226	−5.037	0.0205	−26.65	0.0053	+0.141 / −0.151	−4.996	
				Σ −126.7			0.0442	$\Delta Q_0=\dfrac{-126.7}{2\times0.0442}=+1440\text{ cm}^3/\text{sec}=+1.44$ ℓ/sec	Σ −59.7			0.04458	$\Delta Q_1=\dfrac{-59.7}{2\times0.04458}=+668\text{ cm}^3/\text{sec}=+0.668$ ℓ/sec	Σ −20			0.04422	$\Delta Q_2=\dfrac{-20}{2\times0.04422}=+226\text{ cm}^3/\text{sec}=+0.226$ ℓ/sec	Σ −12.45			0.04476	$\Delta Q_3=\dfrac{-12.45}{2\times0.04476}=+141\text{ cm}^3/\text{sec}=+0.141$ ℓ/sec		
3	KD	1000	150	+7	0.18	+180	0.0257	+0.91	+6.47	0.15	+150	0.0233	+0.362	+6.164	0.141	+141	0.023	+0.352	+6.290	0.145	+145	0.0231	+0.151	+6.30	
	DE	1300	150	+2	0.014	+18.2	0.0091	+0.91	+2.91	0.033	+42.9	0.0147	+0.362	+3.272	0.04	+52	0.0158	+0.352	+3.624	0.047	+61.1	0.0169	+0.151	+3.775	
	EF	1000	150	−6	0.128	−128	0.0113	+0.91	−5.09	0.099	−99	0.0196	+0.362	−4.728	0.082	−82	0.0174	+0.352	−4.376	0.069	−69	0.0157	+0.151	−4.225	
	FK	1300	150	−7	0.18	−234	0.0334	+0.91	−5.755	0.12	−156	0.0270	+0.362	−5.993	0.139	−170.7	0.0285	+0.352	−5.895	0.125	−162.5	0.0276	+0.151 / −0.171	−5.915	
				Σ −163.8			0.0895	$\Delta Q_0=\dfrac{-163.8}{2\times0.0895}=+910\text{ cm}^3/\text{sec}=+0.91$ ℓ/sec	Σ −61.1			0.0846	$\Delta Q_1=\dfrac{-61.1}{2\times0.0846}=+362\text{ cm}^3/\text{sec}=+0.362$ ℓ/sec	Σ −59.7			0.0847	$\Delta Q_2=\dfrac{-59.7}{2\times0.0847}=+352\text{ cm}^3/\text{sec}=+0.352$ ℓ/sec	Σ −25.4			0.0833	$\Delta Q_3=\dfrac{-25.4}{2\times0.0833}=+151\text{ cm}^3/\text{sec}=+0.151$ ℓ/sec		

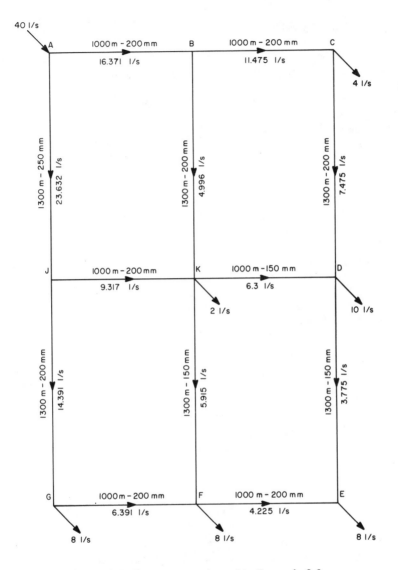

Figure 8.7. Piping network used in Example 8.2.

Flow in the pipe = 1000 ℓ/s.
Assuming a diameter of 700 mm, the head loss is 1 m in 100 m (Figure 8.1).
Total loss of head = 1 x (3200/100) = 32 m.
Piezometric head at A = 200 - 32 = 168 m which is greater than 135 m.
Selecting the next standard size pipe of smaller diameter would have resulted
in too great a head loss.

Design of Pipes AA_1 and AA_2

At A_1 the available head = 168 - (95 + 35) = 38 m.
Loss of head per 100 m = 38 x (100/2000) = 1.9 m.
Flow in pipe AA_1 = 200 ℓ/s.
Assuming a 350 mm pipe, the loss of head is 1.6 m per 100 (Figure 8.1).
At A_2, the available head = 168 - (90 + 35) = 43 m.
Loss of head available per 100 m = 43 x (100/1300) = 3.3 m which is greater
than 1.8 m.

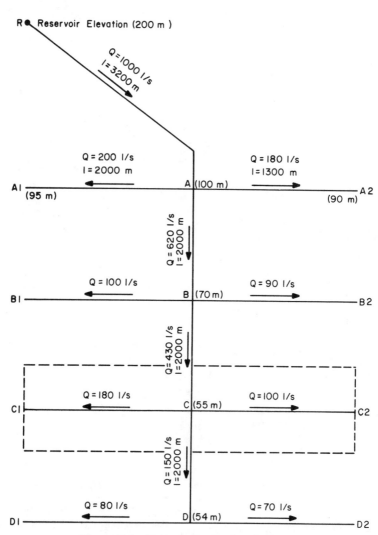

Figure 8.8. Design of a dead-end system.

Flow in the pipe AA_2 = 180 ℓ/s.
Using 300 mm pipe, the loss of head is 3.1 m in 100 m (Figure 8.1).

Design of Pipes BC and CD

At B available head = 168 - (70 + 35) = 60 m.
Allowable loss of head per 100 m = 100 x (60/2000) = 3 m.
Assuming a 600 mm diameter pipe, the loss of head is 0.9 m per 100 m
    (Figure 8.1).
Total loss of head in AB = 0.9 x (2000/100) = 18 m.
Piezometric head at B = 168 - 18 = 150 m = 105 m.
At C the available head = 150 - (55 + 50) = 45 m.
Assuming a 500 mm diameter pipe, the loss of head 1.05 m per 100 m
    (Figure 8.1).
Total loss of head in BC = 1.05 x (2000/100) = 21 m.
Piezometric head at C = 150 - 21 = 129 m = 105 m.

8.2.3 Equivalent Pipe Method

A complicated network of pipes or a number of pipes can be reduced to a single equivalent pipe length or diameter. This will make the computation regarding the flow and pressure easier. In Figure 8.5 a network is given which is to be analyzed by equivalent pipe method (Manning's constant n = 0.012).

(A) Path ABC:
Let Q_1 = 10 ℓ/s;

i.	Pipe AB, 2000 m long, 250 mm diameter,	S = 025%	
		h_f = 0.5 m	
ii.	Pipe BC, 1000 m long, 150 mm diameter,	S = 0.37%	
		h_f = 3.7 m	
	Total	$h_{f_{ABC}}$ = 4.2 m	

Equivalent length of 200 mm pipe = 100 x (4.2/0.08) = 5,250 m.

(B) Path ADC:
Let Q_2 = 6 ℓ/s;

i.	Pipe AD, 1000 m long, 150 mm diameter,	S = 0.14%	
		h_f = 1.4 m	
ii.	Pipe DC, 2000 m long, 200 mm diameter,	S = 0.03%	
		h_f = 0.6 m	
	Total	$h_{f_{ADC}}$ = 2.0 m	

Equivalent length of 150 mm diameter pipe = 100 x (2.0/0.14)
    = 1,450 m.

(C) Equivalent Pipe AC, assume h_f = 2.0 m

    i.   Pipe ABC, 5,250 m long, 200 mm diameter, S = (2.0/52.5)
$$= 0.0381\%$$
$$Q_1 = \ 7 \ \ell/s.$$

    ii.   Pipe ADC, 1,450 m long, 150 mm diameter, S = (2.0/14.5)
$$= 0.138\%$$
$$Q_2 = \ 5.9 \ \ell/s$$
$$\text{Total} \quad Q = \ 12.9 \ \ell/s$$

Equivalent length of 200 mm diameter pipe with
$$Q = \ 12.9 \ \ell/s, \qquad S = 0.16\%$$
$$= \frac{100 \times 2.0}{0.16} = 1,250 \ m$$

Therefore, the network is equivalent to a pipe of 200 mm diameter and 1,250 m length.

9. Water Quality

9.1 INTRODUCTION

The quality of water depends upon its physical, chemical, and biological characteristics. Natural water contains impurities; whereas, pure distilled water is composed of only hydrogen and oxygen. Obtaining a desired quality of water frequently requires treatment. The treatment depends upon the quality of the raw water and the desired quality of treated water.

9.2 IMPURITIES IN WATER

Natural water contains suspended as well as dissolved substances. Frequently these must be either removed or kept within certain limits to make the water potable. The tolerable limit of impurities in water depends upon the purpose for which it is used. Water completely free from suspended or dissolved matter, e.g. distilled water, is unpalatable. Moreover, such a water lacks trace elements necessary for good health. Water for domestic purposes must not contain disease causing organisms. Water used for washing in a laundry or in a textile factory should be free of suspended matter. In brief, water must suit the respective needs of consumers.

The impurities found in water are due to its contact with the air, soil and wastewater from communities and industry. With the help of certain processes it can be purified and made suitable for water supply. These methods are discussed in other chapters. The common impurities found in water are noted in Table 9.1. The types of impurities which may be present in water from various sources are discussed in the following sections.

9.2.1 Meteorological Water

Meteorological water is rain water stored in reservoirs. If the reservoir is properly maintained, water can be safely supplied after disinfection.

9.2.2 Surface Water

Surface water is found in rivers, lakes, and ponds. Substances found in surface water depend upon the watershed. Impurities like clay, organic and

133

Table 9.1. Common Impurities and Their Effects.

Dissolved substances:	
Gases	: Oxygen — promotes corrosion : Carbon dioxide — causes low pH resulting in corrosion : Hydrogen sulfide — foul odor, low pH, corrosion : Nitrogen — : Methane — may cause explosion in contact with air

Inorganic salts	: Dissociates into cations and anions
	: Cations, Ca^{++}, Mg^{++}; cause hardness
	    Na, K^+; increase salt concentration, can import tastes
	    Fe — taste, "red water"
	          corrosion, if > 0.1 mg/ℓ
	    Mn — "black or brown" water
	    H^+ — corrosion, low pH

Anions	: HCO_3^-, $CO_3^=$, OH^-, $SO_4^=$ — alkalinity : Cl^- — taste : F^- — mottled enamel of teeth between 1.5-3 mg/ℓ
Organic compounds	: Color caused due to natural dyes (swamp water) odor and taste due to plant life e.g. algae, or due to certain organics e.g. phenol from industrial waste
Suspended matter	: Mineral matter, silts, clay — cause turbidity : Organic matter (fragments of leaves, grass, etc., — col- loidal suspension — turbidity
Biological life	: Bacteria, viruses — some cause disease : Algae, diatoms — odor, taste, turbidity, color
Protozoa	: Some cause disease

mineral matter, algae, bacteria, and protozoa may be found either in suspended or colloidal form.

Dissolved gases like oxygen, nitrogen, carbon dioxide, methane, and hydrogen sulfide may be present. Organic matter, ammonia, organic acids, chlorides, nitrites, and nitrates may also be found in the dissolved state.

9.2.3 Groundwater

Substances found in groundwater depend upon the characteristics of the underground strata. The water may contain clay, aluminum, iron oxide, silica, and fungi in suspended and colloidal form. Dissolved substances like bicarbonates, carbonates, chlorides, sulfates, and nitrates (salts of magnesium, calcium, sodium, and potassium) may be found. Carbon dioxide, bicarbonate, and   sulfate of iron, sodium chloride, and compounds of manganese may also be present.

9.3 WATER-BORNE DISEASES

Water offers a very suitable environment for the growth of micro-organisms. In the absence of proper care, many disease causing organisms (pathogens) may enter the water and be carried to the users, causing the spread of disease. A communicable disease that can be transmitted through water is called a water-borne disease. Some of the water-borne diseases are cholera, typhoid, paratyphoid, jaundice, bacillary dysentery, amebic dysentery, gastroenteritis and schistosomiases (caused by an animal parasite living in water snails).

9.4 EXAMINATION OF WATER

Water used for domestic purposes must be palatable and free from objectionable impurities. The quality of water can be judged by the following three types of tests: (1) physical, (2) chemical, and (3) bacteriological.

9.4.1 Physical Tests

Several of the constituents used to identify the physical characteristics of water are summarized in Table 9.2 and discussed in the following paragraphs.

(1) Color: Water should be colorless. Presence of color indicates that water contains dissolved and colloidal matter, which is objectionable.

(2) Turbidity: Water becomes turbid when substances like silt, clay, and finely divided organic matter are present. This especially happens with surface waters during rainy seasons. Groundwater is normally clear. Turbidity may be due to organic or inorganic impurities suspended in water. It interferes with the passage of light through the sample of water. With the help of a turbidimeter, one can determine the level of turbidity in a water sample.

(3) Taste: Water should not have a bad taste. Presence of a slight amount of salt may spoil the taste. Certain organic and inorganic impurities, if present even in the slightest amount, may make water unpalatable. Presence of phenol even in the order of 10^{-3} mg/ℓ may be noticeable. Algae, iron, and magnesium salts also impart bad taste to water.

(4) Odor: Any odor in water is objectionable. The intensity of odor can be expressed as a threshold odor number. The threshold odor number of a sample can be obtained by diluting a quantity of the sample to 200 mℓ, so that the odor is barely perceptible in comparison to odor free water. Therefore, if 20 mℓ of the sample is diluted to 200 mℓ, the threshold odor number = 200/20 = 10.

(5) Temperature: Either very hot or very cold water is objectionable. The temperature should be between 7° and 12° C. Information about the

Table 9.2. Drinking Water Standards Proposed by U.S. Public Health Service

	mg/ℓ
Turbidity (silica scale)	10
Color (Platinum or cobalt scale)	20
Lead	0.1
Fluoride	1.5
Arsenic	0.05
Selenium	0.05
Hexavalent chromium	0.05
Nitrate	No upper limit, however, maximum of 10 mg/ℓ as N recommended
Copper	3.0
Iron and manganese together	0.3
Magnesium	125
Zinc	15
Chloride	250
Sulfate	250
Phenolic compounds (as phenol)	0.001
Total solids	500

water temperature is essential, because it effects the properties of water such as viscosity, density, and surface tension. The solubility of chemicals and bacteriological activity are also influenced by temperature.

9.4.2 Chemical Tests

Chemical composition of surface water depends upon the characteristics of the catchment area. Groundwater acquires the characteristics of the soil through which it flows. Ions and compounds frequently measured are listed in Table 9.2. The following tests are conducted to evaluate and control the quality of water:

(1) pH value: The pH value indicates whether water is acidic or alkaline. Water is ionized to hydrated hydrogen ions and hydroxyl ions.

$$H_2O + H_2O \rightleftharpoons H_3O^+ + OH^- \quad \ldots \ldots \quad (9.1)$$

In pure water a 10^{-7} molar concentration of hydrated hydrogen and hydroxyl ions are present at 18° C. When water becomes acidic, the number of hydrated hydrogen ions increases. The H_3O^+ ion concentration in the solution is expressed as the logarithm of its reciprocal, and this is called the pH value of the solution.

$$\text{pH value} = -\log (H_3O^+) = \log -\frac{1}{(H_3O^+)} \quad \dots \quad (9.2)$$

$$= \log \frac{1}{(10^{-7})} = 7 \text{ (for pure water at } 180°C)$$

If the pH value is more than 7, it is alkaline, and if less than 7, acidic. A slight variation in the pH value will mean a large variation in the ion concentration. If the pH value of a particular water is 8, it means that the H_3O^+ ion concentration is 10^{-8}. This water is 10 times as alkaline as a neutral water, which has a H_3O^+ ion concentration of 10^{-7}.

The pH value of water is of great importance. Control of the coagulation process, removal of iron and manganese, and taste and corrosion control are directly related to the pH value.

(2) Carbon dioxide: Excess CO_2 in water is disadvantageous because it can make water corrosive to metals. It can be present in water in the form of carbonates, bicarbonates, or as free CO_2.

(3) Hardness: Salts of calcium and magnesium make the water "hard." Under acidic conditions water can dissolve many of these salts. To a certain degree hardness may be beneficial to health, but when present to a high degree it affects the taste of water and soap consumption in cleaning.

Hardness can be of two types, temporary or permanent. Temporary hardness is due to carbonate and bicarbonates of calcium and magnesium salts. This can be removed simply by boiling. Permanent hardness is due to sulfate, chloride, nitrate and silicate of calcium and magnesium. A combination of these is called total hardness.

Hardness is expressed as mg CaCO3/ℓ of water. It is also expressed in degrees and relative terms (Tables 9.3 and 9.4).

Table 9.3. Definition of Degrees of Hardness

1 degree (English hardness)	= 10 mg $CaCO_3$ in 0.7 ℓ of water
	= 14.3 mg $CaCO_3$/ℓ of water
1 degree (German hardness)	= 10 mg CaO/ℓ of water
1 degree (French hardness)	= 10 mg $CaCO_3$/ℓ of water

Table 9.4. Classification of Waters in Relative Terms of Hardness[1]

Degree of Hardness	Hardness of $CaCO_3$ (mg/ℓ)
Soft	0-55
Slightly hard	56-100
Moderately hard	101-200
Very hard	201-500

(4) Chlorides: Excess chloride in water imparts a bad taste. The chloride concentration should not exceed 250 mg/ℓ; however, there are many areas of the world where it is necessary to exceed this recommended concentration. Exceeding 250 mg/ℓ does not create a health hazard, but the water may be unpalatable to the people unaccustomed to the taste. Previously its presence in water indicated contact with human waste, but now more reliable indicators of pollution have been found.

(5) Iron: Iron in water imparts a bitter taste. It may be present in the form of bicarbonate or sulfate.   Iron is found in groundwater because of the presence of hematite below the ground surface, and iron is soluble in water containing carbonic acid. When water has enough oxygen or when it comes into contact with oxygen, dissolved iron may precipitate as flocs of ferric hydroxide.

Organisms which utilize iron or its compounds cause taste and odor in water and may result in "red water." These organisms are known as "iron bacteria." Water containing iron is unsuitable for most industrial purposes. It cannot be used in laundries, paper mills, film industries and color manufacturing factories.

Removal of iron becomes necessary if it exceeds 0.1 mg/ℓ. Iron content of natural water is rarely more than 10 mg/ℓ but may go up to 50 mg/ℓ.

(6) Manganese: Manganese is generally found in water along with iron. It also spoils the taste when the concentration exceeds 0.5 mg/ℓ. Impounded waters frequently contain manganese, but it is usually mixed with the sludge at the bottom. White clothes turn yellowish when washed in water containing manganese. Its presence in small quantities affects the color of water. The concentration of manganese should be less than 0.1 mg/ℓ.

(7) Oxygen content: Surface water is usually saturated with dissolved oxygen, but the discharge of wastewater into rivers depletes the oxygen content. Groundwater drawn from great depths may be deficient in dissolved oxygen.

There should be a sufficient quantity of dissolved oxygen in water flowing through pipelines, otherwise iron may dissolve in water causing corrosion of pipes. Boiler water should be free from oxygen. From a hygienic point of view, dissolved oxygen in water is not of importance.

(8) Hydrogen sulfide: Its presence will make water unpalatable by producing bad taste and odor. Hydrogen sulfide is found in groundwater due to contact with particular types of geological strata. The reduction process of organic and inorganic substances may also produce hydrogen sulfide. Hydrogen sulfide can be removed by aeration or a combination of aeration and pH adjustment.

9.4.3 Bacteriological Tests

A wide variety of pathogens are found in water (Sec. 9.3). It is difficult to detect their presence by direct analysis as they die quickly. Along with pathogens, organisms of the coliform group (e.g. *fecal streptococci* and *clostridium Welchii)* are present in large numbers in the human intestine. Therefore, presence of coliform organisms in water indicates contamination. The death rate of these organisms is comparatively slower than organisms of the enteric group such as typhoid or paratyphoid. Therefore, the presence of the coliform group of organisms in a water indicates the possibility that the water may have been in contact with organisms of the enteric group. Thus the presence of coliform organisms suggests fecal pollution and indicates that the water is potentially dangerous.

The life span of various organisms serving as indicators of pollution varies. Some of them die in a few days and some survive for a few weeks. This presents some difficulty in the interpretation of the bacteriological tests. However, coliform organisms are considered suitable indicators of pollution because they die in a short time and can be identified easily. *Escherichia coli* of the coliform group are of greater fecal significance and are found in the human and animal intestine.

Bacteriological tests are conducted to determine the presence of coliform organisms in water. Quality of water is determined by the number of coliforms which can be counted in a particular volume of water. Details regarding the bacteriological examination of water are available elsewhere.[2]

9.5 DRINKING WATER SUPPLY STANDARD

Drinking water must be safe and palatable; therefore, it is essential to limit the concentrations of impurities allowed in a water supply. Every water must conform to certain standards promulgated by the public health authorities. Standards are generally not absolute and may vary with the local conditions and the cost of treatment. The U.S. Public Health Service Standards for drinking water are given in Table 9.2 and may serve as a guide.

Samples collected for analysis must be representative. Tests can be conducted daily, weekly, or monthly. This will depend upon whether the quality is uniform or varies considerably during a week or month.

The frequency of sample collection for bacteriological analysis specified by the U.S. Public Health Service is given in Table 9.5. Drinking water must not contain more than a single organism in 100 mℓ.

Table 9.5. Frequency of Bacteriological Sample Collection

Population Served	Minimum Number of Samples Per Month
2,500 or less	1
10,000	7
20,000	25
100,000	100
1,000,000	300
2,000,000	390
5,000,000	500

10.   Water Treatment

10.1 GENERAL

The impurities in water must be below certain limits or be removed before using the water for domestic or industrial purposes. This means that with the exception of unusual situations water must be treated. The degree of water treatment will depend upon the quality of raw water and how it is to be used. Its quality must conform to a certain standard. Groundwater, in general, is free from suspended matter and microorganisms; therefore, it can be safely supplied after disinfection. Sometimes groundwater contains high concentrations of calcium and magnesium salts (hardness) which may have to be removed. If hardness is not removed, consumption of soap will be high and scaling will occur in boilers and hot water heaters. Iron and manganese in concentrations greater than 0.5 mg/ℓ must also be removed. Excess carbon dioxide should also be removed to prevent corrosive water.

If the catchment area of an impounded reservoir is free of human habitation and proper sanitary control is exercised so that wastewater does not discharge into the reservoir, water from the reservoir can be supplied to a community after disinfection. A large surface area and a long detention time will help the self purification process in improving the quality of water but will increase evaporation losses.

Surface water contains suspended and dissolved impurities and is generally contaminated. Therefore, it must be properly treated before it can be supplied to a community.

10.2 TREATMENT PROCESSES

The unit operations (steps in treatment) listed in Table 10.1 are employed in various combinations to produce industrial and culinary water supplies. Depending upon the raw water quality, all or a combination of operations may be used. A typical combination of unit operations is shown in Figure 10.1.

10.3 SCREEN

Strainers in the intake wells or intake pipes prevent large floating matter from entering a water treatment plant. Fine floating matter is usually controlled using screens with small openings 0.95 cm (mesh size 3/8 inch). In some situations coarse screens with openings between 2.5 and 7.5 cm preceding the finer screens may also be used.

141

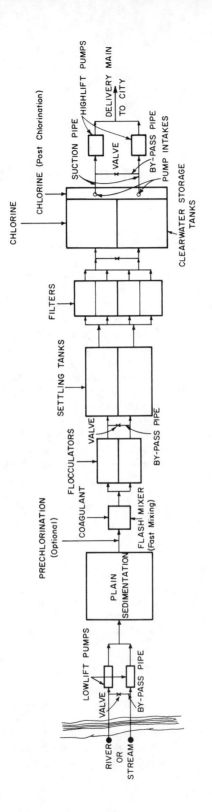

Figure 10.1. Flow diagram of a water treatment plant.

Table 10.1. Unit Operations Employed in Water Treatment

(1)	Gas Transfer	: Removal or addition of gases such as oxygen and carbon dioxide.
(2)	Screen	: For the removal of floating matter.
(3)	Storage	: Impounded reservoirs and lakes to store water.
(4)	Presedimentation	: Sedimentation tank with long detention time for the removal of suspended matter in a highly turbid water.
(5)	Coagulation and Flocculation	: For encouraging the nonsettleable particles to form flocs so that they may be removed by settling.
(6)	Settling	: Settling tank with short detention time for the removal of suspended flocs.
(7)	Filtration	: For the removal of finely divided particles, suspended flocs, and microorganisms.
(8)	Disinfection	: For killing microorganisms.

To remove certain chemicals or taste or odor from water, the following methods are applied:

(1)	Adsorption	: For the removal of taste and odor.
(2)	Chemical Precipitation	: For the removal of dissolved substances, such as iron, manganese, and hardness.
(3)	Ion Exchange	: Exchange of ions of certain salts present in water for ions present in an exchange medium.

10.4 STORAGE

Water stored in reservoirs frequently can be supplied to a community after disinfection. Most of the suspended particles are removed and about 90 percent of the pathogens die after a detention time of one week. Algae control in reservoirs is frequently a problem in the summer months. Algae create taste and odor problems and can interfere with filter operation. Microstrainers have been successfully employed in such cases.

10.5 PRESEDIMENTATION

Large earthen tanks with a long detention time may be useful with highly turbid waters (10,000 mg/ℓ and above). The use of presedimentation basins should not be used to justify higher loading rates for settling tanks and filters. Heavy suspended particles are quickly removed and in many cases can reduce the turbidity to about 1 mg/ℓ. Algae control during the warm seasons of the year can require considerable effort.

11. Sedimentation

11.1 TYPES OF SEDIMENTATION

Sedimentation is the separation of solid particles from a suspension by the force of gravity, and the suspension is separated into clarified liquid and concentrated suspension. The terms clarification and thickening are also used to describe this process. When the objective is a clarified liquid, then the term clarification is used. When the objective is a concentrated suspension below the clarified liquid, the term thickening is used.

The sedimentation process is used to remove from water discrete particles, flocculated matter, and precipitates formed during various water treatment operations. This process has been divided into four categories depending on the concentration of the suspension and the flocculating properties of the particles. The categories are discrete settling, flocculent settling, zone settling, and compression settling. The first two types deal with dilute suspensions, the particles of the first type being discrete and of the second type being flocculent. In zone settling the mass of particles which are flocculent materials in suspensions of intermediate concentration settle as a whole, because the particles are so close that interparticle forces keep them in a fixed position relative to one another. When the concentration is so high that the particles are in physical contact with each other and are supported by the compacting mass, compression takes place. Liquid in the suspension mass is slowly displaced when compression occurs.

11.2 SETTLING OF DISCRETE PARTICLES

A discrete particle moving vertically downward through a quiescent fluid will accelerate until the frictional resistance of drag of the fluid equals the impelling force acting upon the particle, and after that the particle will settle with a uniform velocity.

The impelling force F_i equals the weight of the particle in the suspended fluid,

$$F_i = (D_p - D) gV \quad \ldots \ldots \ldots \ldots (11.1)$$

145

where D_p and D are mass densities of the particle and the fluid respectively and V is the volume of the particle.

The drag or the frictional resistance F_d is a function of dynamic viscosity, μ, density of the fluid, D, settling velocity, V_s, and the characteristic diameter, d, of the particle,

$$F_d = f(V_s, d, D, \mu)$$

By dimensional analysis and experimentation the expression for the drag force

$$F_d = \tfrac{1}{2}C_d A_c DV_s^2, \quad \cdots \cdots \cdots (11.2)$$

was developed in which C_d is the Newton drag coefficient and A_c is the cross-sectional area of projection of the particle at right angles to the direction of settling.

For spherical particles, C_d depends upon Reynold's number and is given by

$$C_d = \frac{24}{R_e} + \frac{3}{\sqrt{R_e}} + 0.34 \quad \cdots \cdots \cdots (11.3)$$

Equation 11.3 applies in the transition and turbulent boundary regions (up to $R_e = 1 \times 10^4$). Equating the values of F_i and F_d and solving we get

$$V_s = \sqrt{\frac{2g(D_p - D)V}{C_d \, D \, A_c}} \quad \cdots \cdots \cdots (11.4)$$

For spherical particles $V = \dfrac{\pi d^3}{6}$ and $A_c = \dfrac{\pi d^2}{4}$

Therefore $\dfrac{V}{A_c} = \dfrac{2}{3} d$.

Consequently the settling velocity of the particle

$$V_s = \sqrt{\frac{4g}{3C_d} \frac{(D_p - D) d}{D}} \quad \cdots \cdots \cdots (11.5)$$

For the laminar region $(R_e < 1.0)$ $C_d = \dfrac{24}{R_e}$ and $R_e = \dfrac{DV_s d}{\mu}$. Hence, from Equation 11.5

$$V_s = \frac{g(D_p - D) d^2}{18\mu} \quad \cdots \cdots \cdots (11.6)$$

Equation 11.6 is known as Stoke's law.

The settling velocity of particles can be found from Equations 11.5 and 11.6 if the particles are spherical and do not change size and shape during settling. But actually particles do not remain discrete during settling nor are they truly spherical. Normally the suspended matter has a large surface area in comparison to its volume. This causes an increased resistance to the particles during settling. However, these equations may be used for checking the design of settling tanks.

Stoke's law is applicable for small diameter particles (up to 0.1 cm diameter) and Reynold's numbers ranging between 10^{-4} and 1.0. In the transition region where R_e = 1 to 2000, Equation 11.5 can be used. Sand particles and heavy flocs normally settle in the transition region. The overflow rate in settling tanks is generally between 0.85 and 1.7 m^3/m^2-hour. Corresponding to these overflow rates, theoretically discrete particles with a specific gravity of 1.2^* less than about 0.005 cm in diameter cannot be removed. But in settling tanks even smaller diameter particles may be removed due to agglomeration.

When the concentration of the suspended particles is large, particles cannot settle freely. In this case there is an upward displacement of the fluid, which hinders settling. Theoretically, due to hindered settling, velocities may be reduced by about 99 percent of their normal rate if the volumetric concentration of solids is about 0.22 percent. This value corresponds to about 6000 mg/ℓ for silt and sand. Water normally does not have suspensions in such high concentrations, except river waters during rainy seasons. Certain proprietary devices incorporating flocculation and sedimentation take advantage of hindered settling by developing a thick sludge blanket through which the water flows trapping suspended particles. These devices are discussed in the chapter on coagulation.

11.3 SEDIMENTATION IN AN IDEAL TANK[1]

An ideal settling tank consists of the following (Figure 11.1):
1.   An inlet zone where water is distributed along the cross section.
2.   A settling zone which removes suspended particles and is in a quiescent state.
3.   A sludge zone which collects settled particles.
4.   An outlet zone through which water along with nonsettleable particles is carried outside the tank.

Water flows in a rectangular tank with a horizontal velocity V_0 and the suspended particles settle with a velocity V_s. $Q(m^3/hr)$ is the rate of flow of water and V is the volume (m^3) of the tank of dimension ℓbh (Figure 11.1a). Therefore,

$$V_o = \frac{Q}{bh} \ (m/hr) \ \ldots \ldots \ldots \ldots \ldots \ldots \ (11.7)$$

The detention time (theoretical time for which the water is detained in the tank) is

---

*Specific gravities of suspended particles range from 1.03 to 2.65.

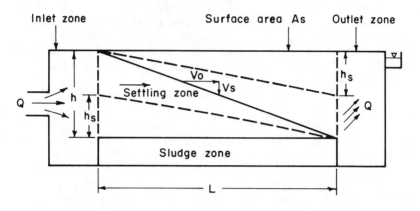

(a.) Cross section of rectangular tank

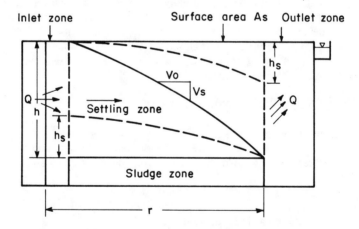

(b.) Cross section of circular tank

Figure 11.1. Settling of a discrete particle in idealized tanks—horizontal flow type.

$$t = \frac{V}{Q} = \frac{\ell bh}{Q} = \frac{\ell}{V_o}(hr) \quad \ldots \ldots \ldots \ldots (11.8)$$

and settling time (theoretical time required for a particle to settle a distance h)

$$t_s = \frac{h}{V_s} \text{ (hours)} \quad \dots \dots \dots \dots \dots \text{(11.9)}$$

where h = depth of water through which the particle has to fall and t_s = time required for particle to settle a distance h. For any particle to be removed, it is necessary that the suspended particle enter the sludge zone just as water is leaving the tank. Therefore, time of flow = settling time;

$$t = t_s = \frac{\ell}{V_o} = \frac{h}{V_s}$$

Therefore
$$V_s = \frac{Vh}{\ell} = \frac{Q}{b\ell} = \frac{Q}{A_s} = S_o \text{ (m/hr)} \quad \dots \dots \text{(11.10)}$$

where A_s = bℓ, the surface area of the tank. In a circular tank of volume $V(m^3)$ having radial flow (Figure 11.1b), the velocity of flow

$$V_o = \frac{Q}{A} = \frac{Q}{\pi rh} \text{ (m/hr)} \quad \dots \dots \dots \dots \text{(11.11)}$$

in which
Q = rate of flow, m^3/hr
A = average area = $(2\pi rh)/2 = \pi rh$
r = radius of the tank
h = depth of water through which particle has to fall

$$\text{Detention time} = t = \frac{V}{Q} = \frac{\pi r^2 h}{Q} = \frac{r}{V_o} \quad \dots \dots \text{(11.12)}$$

$$\text{Settling time} = t_s = \frac{h}{V_s} = t = \frac{\pi r^2 h}{Q} \quad \dots \dots \text{(11.13)}$$

$$\text{Therefore, } V_s = \frac{Q}{\pi r^2} = \frac{Q}{A_s} = S_o \text{ (m/hr)} \quad \dots \dots \text{(11.14)}$$

From Equations 11.10 and 11.14 it is evident that the settling velocity mainly depends upon the surface area and consequently on the surface loading, S_o, (overflow rate) of the tank. The larger the surface area, the smaller will be the settling velocity resulting in the removal of smaller particles.

Since $S_o = V_s = h/t_s$, a greater depth reduces the efficiency of the tank, and a longer detention time increases the efficiency. Tanks are designed considering the surface loading (overflow rate) and the detention time. Tanks have sometimes been subdivided horizontally (Figure 11.5c) to achieve small surface loadings by having a larger surface available for settling. But these tanks have not been popularly used because it is difficult to remove sludge.

All particles having a settling velocity greater than V_s theoretically can be removed. Even those particles whose settling velocity is less than V_s can be removed in horizontal flow tanks if they enter the tank at a height $h_s = V_s t$ (Figure 11.1). In vertical flow tanks such particles cannot be removed if the settling velocity is less than V_s.

11.4 SETTLING ANALYSIS

A wide variety of particle sizes may be found in a suspension, and it is difficult to calculate theoretically the settling velocity of the discrete particles from Equation 11.5. Thus, it may not be possible to predict the percentage removal of particles in a settling tank. However, by conducting a settling analysis, the overall removal of the suspended particles can be estimated for any overflow rate.

A settling column of 200 mm diameter (not less than 130 mm) is constructed with sampling points at different heights (Figure 11.2). The column is filled with the suspension having uniform concentration (well mixed) throughout the depth. The suspension is allowed to settle under quiescent conditions. Samples are taken from a particular depth at different intervals of time, and the concentration of particles in each case is determined. At any sampling depth only those particles whose settling velocities are less than h/t will be found in suspension, where h = the sampling depth, and t is the settling time. With the help of these values a frequency distribution curve, as given in Figure 11.3, can be plotted.

For an overflow rate of $Q = V_s A$, the particles having settling velocities greater than V_s will be completely removed in horizontal flow tanks. The concentration of such particles is $1-C_o$. The proportion of particles having settling velocities of less than V_s will be V/V_s. Therefore the concentration of particles removed will be

$$\frac{1}{V_s} \int_0^{C_o} V dc$$

Thus, total removal from the suspension will be

$$C = (1 - C_o) + \frac{1}{V_s} \int_0^{C_o} V dc \quad . . . . . . (11.15)$$

The second term in Equation 11.15 is determined graphically as shown in Figure 11.3.

In vertical flow tanks the total removal will be $1-C_o$. The removal can be more if particles agglomerate.

11.5 DETENTION TIME

In highly turbid river water containing sand, silt, and clay the percentage removal of suspended matter may be about 50 percent in 1 hour and about 70 percent in 2 hours. The rate of removal of solids after 2 hours of detention time is small. Presedimentation tanks used for highly turbid water may have a detention time of about 8 hours. The performance of settling tanks generally used in water treatment is dependent upon the coagulation and other chemical processes which precede it. Flocs have a

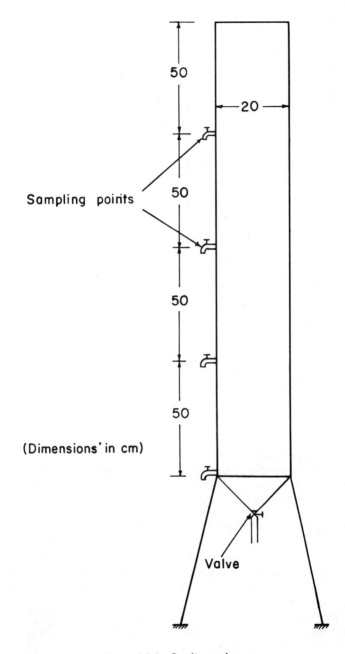

Figure 11.2. Settling column.

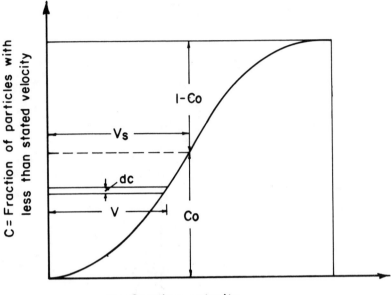

Figure 11.3. Settling analysis curve for discrete particles.

tendency to change their size and shape due to coalescence with other flocs, causing flocculent settling. Estimates of the detention times in such tanks can be determined in the laboratory (Section 11.4). In general the detention time in water supply settling tanks is between 2 and 3 hours.

11.6 FLOCCULENT SETTLING

Flocculent settling was mentioned above in Section 11.5, and when it occurs, as is the case in water treatment, the settling velocity is increased and a plot of the settling characteristics results in a curvilinear relationship as shown in Figure 11.4.

When designing a settling tank, both the overflow rate (surface loading) and detention time should be considered. Flocculent settling has not been formulated; therefore, it is advisable to conduct settling analyses in the laboratory.

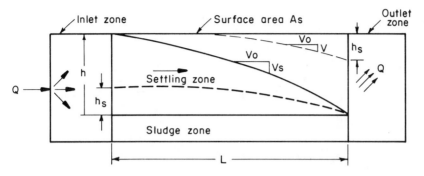

Figure 11.4. Settling path in an idean settling tank—flocculated settling.

11.7 TYPES OF SETTLING TANKS

Settling tanks can be either for horizontal flow or vertical flow (Figures 11.5, 11.6, 11.7). These tanks can be circular, rectangular, or square in plan. The depth may be between 2 to 5 m, the average being 3 m. Rectangular tanks may be up to 30 m long and 10 m wide. Dimensions of the mechanical scrapers also influence the size of the tank. Square tanks may be about 25 m long. The slope at the bottom may be between 2 and 6 percent. The slope of sludge hoppers lies between 1:1 to 2:1 (vertical:horizontal) but 1:1 is more common.

The sludge collected at the bottom or in the hopper is removed either by flushing it into a sump or collecting it into a hopper and then withdrawing it by gravity or pumping. It can also be removed under hydrostatic pressure due to water in the settling tank.

Sludge scrapers may be connected to rotating arms or to endless chains. Scum may also be removed by another attachment to the sludge scraping mechanism near the surface. When a number of rectangular tanks are placed side by side, one scraper may clean all the tanks one after the other by shifting its position sideways. The velocity of the scraper is about 0.3 m/sec. Proper inlet and outlet designs are essential for achieving high efficiency. The function of the inlet is to distribute the inflow uniformly over the entire cross section, and there should be no eddies or cross currents in settling tanks. The velocity of water near the bottom should not scour the deposited sludge.

Various forms of inlets are used so that the energy of the incoming water is minimized. A few of them are shown in Figure 11.8.

When the density of the inflow is greater than the contents of the tank, density currents are sometimes formed, i.e. water tends to move downwards and flow along the bottom. To prevent such phenomena, various types of outlets are provided at the end of the tank for taking out clear water (Figure 11.8). Outlet channels may have weirs on one side or on both sides. For uniform distribution, notches are provided in the weir. The weir should have a length adequate to prevent currents that will disturb the settled solids. Therefore, weir loading, i.e. the flow per unit length (m^3/hr-m), should not exceed 25 m^3/m-hr.

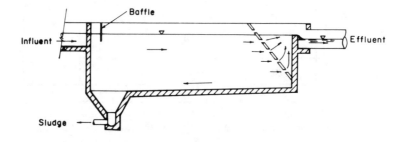

(a.) Without mechanical cleaning: for cleaning, the tank must be put out of operation

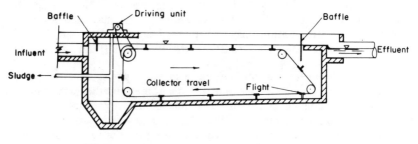

(b.) Mechanically cleaned

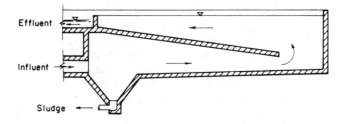

(c.) With single tray (tray increases total surface area)

Figure 11.5. Rectangular settling tank—longitudinal flow.

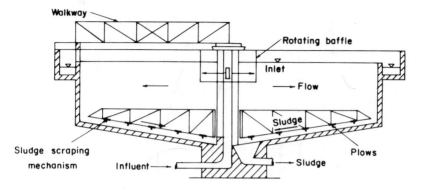

Figure 11.6. Circular settling tank with sludge scraping mechanism.

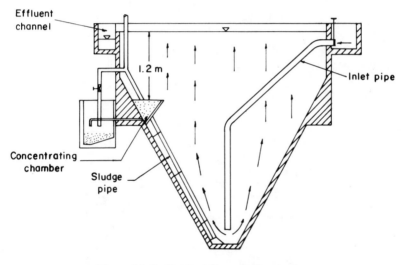

Figure 11.7. Vertical flow settling tank.

11.7.1 Tube Settler

To improve the performance of existing and new settling basins, tube settlers may be employed. Tube settlers are installed in both rectangular and circular clarifiers and can be used with upflow clarifiers with solids contact.

Tube settlers are available in two basic configurations: essentially horizontal tubes and steeply inclined ones. The horizontal tubes are operated in conjunction with the filter that follows sedimentation. The tubes become filled with solids and are completely drained with each filter backwash. Horizontal tube settlers are used in small plants with a capacity of less than 1 mgd

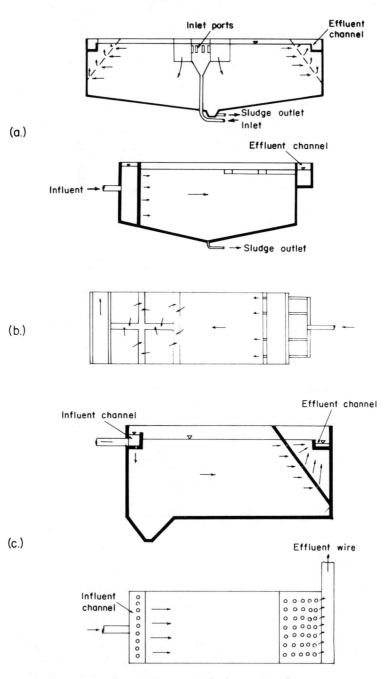

Figure 11.8. Types of inlets and outlets.

(3,785 m³/day) or in package plants. Steeply inclined tube settlers continuously remove sludge due to the flow pattern established. Because of the shallow depth of the inclined tube settlers and continuous sludge removal, the size of installation is unlimited.

Success has been reported with tube settlers, and the application of these devices should be considered.[2] At this time adequate data do not exist to make efficiency and economic analyses of the various proprietary units available. Typical applications of steeply inclined tube settlers are shown in Figures 11.9 and 11.10.

Example 11.1

Find the following:
(a)   The settling velocity in water at 20°C of a spherical particle 4.5×10^{-3} cm in diameter and having a specific gravity = 2.65.
(b)   The rising velocity of the particle in (a) if its specific gravity is 0.85.
(c)   The settling velocity at 20°C of a spherical particle 0.12 cm in diameter and having specific gravity = 2.65.

Figure 11.9. Chevron tube settlers in vertical precipitator. (Courtesy of Permutit Company Inc.)

Figure 11.10. Chevron tube settlers in horizontal precipitator. (Courtesy of Permutit Company Inc.)

Solution:

(a) $V_s = \dfrac{981\,(2.65 - 1)\,(4.5 \times 10^{-3})^2}{18 \times 1.009 \times 10^{-2}} = 0.18 \text{ cm/s}$

$R_e = \dfrac{VdD}{\mu} = \dfrac{0.18 \times 4.5 \times 10^{-3} \times 1}{1.009 \times 10^{-2}} = 0.08 < 1$

Hence Stoke's law is applicable.

(b) $V_s = \dfrac{981\,(0.8 - 1)\,(4.5 \times 10^{-3})^2}{18 \times 1.009 \times 10^{-2}} = -0.022 \text{ cm/s}$

The negative sign indicates that the particle is rising instead of settling.

$R_e = \dfrac{0.22\,(4.5 \times 10^{-1}) \times 1}{1.009 \times 10^{-2}} = 0.0098 < 1$

Hence Stoke's law is applicable.

(c) $V_s = \dfrac{981\,(2.64 - 1)\,(0.12)}{\cdot 18 \times 1.3097 \times 10^{-2}} = 824 \text{ cm/s}$

$R_e = \dfrac{824 \times 0.12}{1.3097 \times 10^{-2}} = 7550 > 1$

Hence Stoke's law is not applicable.

Using $\quad V_s = \sqrt{\dfrac{4}{3}\dfrac{g}{C_D}\dfrac{(D_p - D)}{D}d}$

We get $\quad V_s = 21$ cm/s (approx.) by trial.

Example 11.2

Analysis of a dilute suspension containing discrete particles is performed in a settling column. Data collected from samples taken at a depth of 2 m are given below:

Sampling Time (min)	Portion of Particles in Sample
0.7	0.60
1.2	0.50
2.3	0.35
4.6	0.15
6.3	0.08
8.8	0.04

(a) Find the overall removal if the overflow rate of the settling basin is 0.03 m/s (1.80 m/min).
(b) Find the overall removal if a tray is installed at mid-depth.
(c) Determine the increase in efficiency.

Solution:

Time (min)	Settling Velocity $(m/min) = \dfrac{2\ m}{time}$	Portion of Particles with Velocity Less Than Indicated C
0.7	2.86	0.60
1.2	1.66	0.50
2.3	0.87	0.35
4.6	0.43	0.15
6.3	0.32	0.09
8.8	0.23	0.04

(a) $\quad C = (1 - C_o) + \dfrac{1}{V_s}\displaystyle\int_0^{C_o} V dC$

From Figure 11.11, graphical integration yields $\displaystyle\int_0^{C_o = 0.49} V dC = 0.349$

Therefore $\quad C = (1 - 0.49) + 1/1.8\,(0.349)$
$= 0.51 + 0.188 = 0.698$
$= 69.8\%$ removal

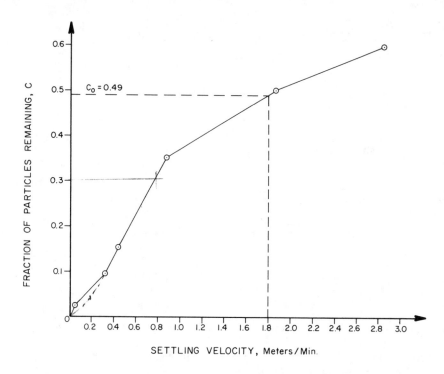

Figure 11.11. Graphical integration of settling column data for Example 11.2.

(b)   Introduction of a tray doubles the surface area, hence

$$V_s = \frac{Q}{2A} = \frac{1.8}{2} = 0.9 \text{ m/min}$$

From Figure 11.11 $C_0 = 0.35$ corresponding to $V_s = 0.9$ m/min

By graphical integration, $\displaystyle\int_0^{C_0=0.35} V dC = 0.153$

Therefore, $C = (1 - 0.3505) + 1/0.9 \times 0.153$
$= 0.8155 = 82\%$ removal

(c)   Increase in efficiency of removal due to tray = 82 - 69.8 = 12.2%

Example 11.3

A flocculent suspension is placed in a settling column (Figure 11.2) and the following results are obtained (Table 11-1). Find the relationship between detention time, overflow rate, and suspended solids removal.

Table 11.1. Results of Settling Column Test in Example 11.3

Time (min)	Conc. of Susp. Solids at the Beginning (mg/ℓ)	Suspended Solids (mg/ℓ) in Sample at Various Depths (m)			% Removal at Various Depths (m)		
		0.5	1.0	2.0	0.5	1.0	2.0
20	1200	800	950	1025	$\dfrac{1200-800}{1200} \times 100$ $= 33.3$	$\dfrac{250}{1200} \times 100$ $= 21$	$\dfrac{175}{1200} \times 100$ $= 15$
40		550	740	800	54.1	38.33	33.33
60		350	500	700	70.8	58.33	41.66
80		200	325	610	83.33	72.91	49.16
100		160	180	460	86.66	85.0	61.66
120		-	-	320	-	-	73.33

Solution:

All of the percentage removals shown in Table 11.2 are plotted on linear graph paper versus sampling depths and times (Figure 11.12a). Curves connecting points of equal removal are drawn by estimation. These curves indicate the settling rate of a certain percentage of suspended solids removed.

Figure 11.12a indicates that 20 percent of the suspended solids are completely removed in a depth of 2 m in 25 min. Therefore, the settling velocity of these particles is 2/25 x 60 = 4.8 m/hr. All particles having a velocity equal to or greater than 4.8 m/hr will be removed in an ideal settling tank having an overflow rate of 4.8 m³/m²-hr. Particles having a settling velocity less than 4.8 m/hr will be removed in the ratio v/4.8 or in the ratio of the average depth settled to total depth of 2.0 m.

The average depth to which the next 10 percent (between 20 and 30 percent) of the particles has settled is 1.45 m (Figure 11.12a), and therefore the percentage removal of this fraction is 1.45/2 x 10% = 7.25 percent.

Similarly the removal of suspended solids has been calculated for the subsequent 10 percent ranges (i.e. 30 to 40 percent, 40 to 50 percent, etc.).

$$\text{Thus the total removal} = \left(20 + \frac{1.45}{2} + \frac{0.65}{2} \times 10 + \frac{0.349}{2} \times 10\right)$$

$$= (20 + 7.25 + 3.25 + 1.8) = 35.3\%$$

This removal takes place at an overflow rate of 4.8 m³/m²-hr and a detention time of 25 min. Similarly the total removal corresponding to other values of overflow rates and detention time can be calculated. From these values (Table 11.2) curves are drawn from which the design criteria can be ascertained (Figure 11.12b).

Table 11.2. Percent Removals of Suspended Solids in Example 11.3

Time (min) to Settle through 2 m	Over Flow Rate (m³/m²-hr) or Settling Velocity (m/hr)	% Removal of Suspended Solids
25	4.8	35.3
32	3.75	40.8
57	2.11	55.84
81	1.48	59.22
96	1.25	68.63
115	1.04	70

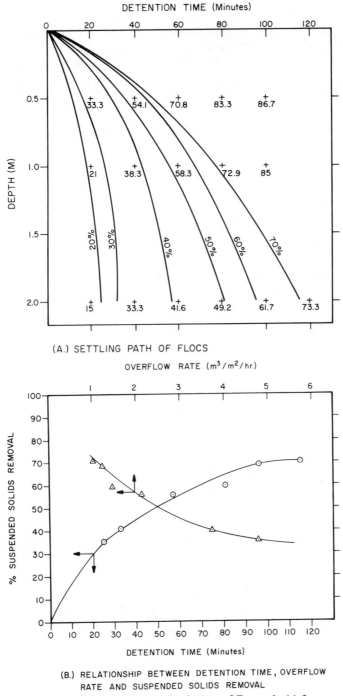

DETENTION TIME (Minutes)

(A.) SETTLING PATH OF FLOCS

OVERFLOW RATE (m³/m²/hr.)

% SUSPENDED SOLIDS REMOVAL

DETENTION TIME (Minutes)

(B.) RELATIONSHIP BETWEEN DETENTION TIME, OVERFLOW
RATE AND SUSPENDED SOLIDS REMOVAL

Figure 11.12. Graphical solution of Example 11.3.

Figure 11.12b indicates that for 70 percent removal the detention time is 105 min and the corresponding overflow rate is 1.12 m^3/m^2-hr. To account for short circuiting, eddies, turbulence, and disturbances at inlet and outlet, it is suggested that the overflow rate be decreased by a factor of 1.5 (1.25 to 1.75) and to increase the detention time by a factor of 1.75 (1.5 to 2.0).

12. Coagulation

12.1 HISTORICAL BACKGROUND OF COAGULATION

The process of coagulation has been used for centuries by mankind to clarify drinking water. Narmali seed or "Clearing Nut" was used for many centuries in the rural villages of Southern India and Ceylon, long before the more modern coagulant and synthetic polyelectrolytes were used.[1] Turbidity and organic color were the only impurities of concern until recently when the rise in the standard of living and rapid urban development created a steady increase in water consumption. The water supply industry has been required to supply large quantities of water and at the same time to develop the techniques for maintaining high quality. A large variety of new substances has appeared in the raw water due to the advanced development of industries and agriculture, and these substances have intensified the problem of water treatment.

Products such as surfactants, herbicides, pesticides, and a large number of organic and inorganic pollutants introduced by industrial processes must now be removed. The coagulation process has been employed for almost a century to remove suspended impurities from waters. Coagulation still remains one of the most important methods of water treatment.

Perhaps the most dramatic development with coagulation in the recent years is the use of new coagulants such as the polyelectrolytes. Polyelectrolytes have been used as a coagulant and as a coagulant aid to increase the efficiency of the removal of turbidity and color and also to remove some of the new pollutants not affected by conventional coagulants.

Inorganic coagulants are used most frequently in water treatment practice. Aluminum sulfate has been and remains the most common coagulant for water purification. Iron salts have been used limitedly and principally in research investigations.

12.1.1 Terminology

Confusion exists over the meaning of *coagulation* and *flocculation*. In the overall process of coagulation-flocculation there is a cause and effect relationship. Reactions of the type encountered in water treatment practice, destabilization and particle collision opportunity can be considered as the *cause*. Aggregation of the destabilized, colliding particles is then the *effect*.

165

Destabilization and particle collisions are independent variables, and aggregation is the dependent variable. To remove colloidal matter both destabilization and particle collision must be provided. Destabilization is considered as the physiochemical change which accompanies the addition of chemicals and is the action which allows particles to adhere to one another. Particle collision opportunities occur as a result of relative particle transport. Increasing the likelihood of collision can be accomplished in a number of ways including the use of paddles, baffled basins, hydraulic jets and turbulence in pipes and channels. Coagulation describes the overall process of particle aggregation including both particle destabilization and particle transportation. Flocculation describes only the transport process.

12.2 DEFINITIONS

Colloids are discrete particles which can remain in suspension in a dispersion medium. To remove them from suspension, it is necessary to coagulate small particles into larger ones. The size of colloidal particles normally varies between 1 and 10 $m\mu$ (1 $m\mu = 10^{-6}$ mm). The small size of the colloid imparts properties which are very different from coarse particles. Colloids are very small, but the surface area is very great relative to the mass of the particle. This can be realized from the fact that if one cubic centimeter is packed with cubical colloids of 10^{-6} cm, then the surface area is 6×10^2 sq m; whereas, the surface area of 1 cc cube is 6 sq cm. Thus, surface properties predominate and are discussed in the following paragraphs.

12.3 COLLOIDAL PROPERTIES

12.3.1 Adsorption

Colloids have a tendency to concentrate substances at their surface, and this collection of particles is known as adsorption. But colloids indicate preference for one ion over another with adsorption depending upon the substance. Adsorption also depends upon concentration and temperature.

12.3.2 Electrokinetic Properties

Colloidal particles are normally electrically charged. According to the nature of the colloid, the charge may vary in magnitude, and it may be positive or negative. For the removal of colloids from a liquid, knowledge of this property is of great importance.

If a D.C. current is passed through a colloidal suspension, particles migrate towards the pole of opposite charge, this phenomena being known as electrophoresis. The charge on a colloid is determined by this method. Metallic oxides (e.g. iron, aluminum) are mostly positively charged, non-metallic oxide, metallic sulfides, clay, organic color and most proteins are usually negatively charged in water.

There are two types of solid colloidal dispersions in a liquid, hydrophobic (solvent-hating) and the hydrophilic (solvent-loving). Colloids usually encountered in the treatment of water and wastewater are not purely hydrophobic. Metallic oxide and non-metallic oxide are hydrophobic colloids. Soap, detergents, soluble starch, and soluble proteins are examples of hydrophilic colloids. Stability of these colloids depends upon the hydration and the electric charge on their surface. The electric charge influences the stability of hydrophobic colloids. Adsorption helps in increasing the charge on the colloids, depending upon the valence and the number of ions adsorbed.

A double layer is formed around a colloidal particle (Figure 12.1). The layer to which the stabilizing ions are adsorbed is called the inner fixed layer

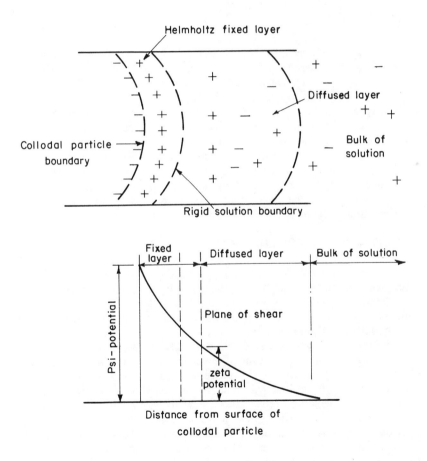

Figure 12.1. Electrochemical behavior at the surface of colloidal particle.

or Helmholtz fixed layer. Around the first layer a diffuse second layer is formed by ions of opposite charge. The psi-potential is the electrical potential between the interface of the colloid and the bulk of the solution. The zeta-potential is the electrical potential between the rigid solution boundary and the bulk of the solution and can be measured for colloidal particles. The rigid solution boundary is defined as the boundary between the liquid which is immovably attached to the colloidal surface and the body of the solution. For dilute solutions of electrolyte and low potentials, the psi-potential and the zeta-potential almost coincide. It is not possible to measure psi-potential, but zeta-potential can be determined. Zeta-potential is defined by the following equation:

$$Z = \frac{4\pi \, qt}{D} \quad \cdots \cdots \cdots \cdots \cdots \quad (12.1)$$

in which

Z = zeta potential
q = charge difference between the particle and the medium
t = thickness of the layer around the particle through which the charge difference is effective
D = dielectric constant of medium

Instability of colloidal particles is caused by Brownian motion and Van der Waals forces of attraction. Due to Brownian motion, particles come close to each other, so that the influence of interacting forces may become effective. The two forces are additive for a small distance and their value becomes high for colloidal particles. Coagulation can be achieved by lowering the zeta-potential. It may be lowered by adding ions of opposite charge. If a diffused part of the electrical double layer is reduced by increasing the ion concentration in the solution, then the zeta-potential is also lowered.

12.4 COAGULATION OF COLLOIDS

Hydrophobic colloids are removed with the help of electrolytes. In the initial phase of coagulation, hydrolysis, crystallization, and compensation of the negative charge on the colloidal impurities takes place, resulting in the formation of microfloc. This molecular coagulation in the initial phase is known as perikinetic coagulation.

Perikinetic coagulation takes place quickly and is completed during the time required for thorough mixing of the coagulant and the liquid. This is the most important phase because this phase determines which substance and to what extent it will be removed from the liquid. The second stage is called orthokinetic coagulation (coagulation due to movement), in which the microflocs agglomerate with other flocs and settle. Stability of hydrophilic colloids depends largely upon their affinity for the liquid rather than the electrical charge on their surface. Due to this, it becomes difficult to remove these colloids from the liquid medium. When hydrophilic colloids serve as a protective layer for the hydrophobic colloids, the coagulation process is hindered. In such a case, the amount of coagulant required will be high. The affinity for the bound water can be reduced by adding salts in high concentration. This method is known as salting out, and sulfate, chloride, nitrate, and iodide anions are listed in descending order of effectiveness.

12.5 CHEMICAL COAGULATION

The most common electrolyte used for removing turbidity, color, taste, odor, bacteria and charge are compounds of iron and aluminum. These compounds when added to water ionize to form cations and anions of high valence.

These ions react with the hydroxyl ion to give colloidal hydrous oxides (hydroxides) which are positively charged. This hydroxide in turn neutralizes the charge on negative colloids and helps in the coagulation. The excess colloidal metallic hydroxide is neutralized by the negatively charged sulfate ions resulting in precipitation.

12.5.1 Coagulants in Common Use

Alum and Lime

If alkalinity is not present in water in adequate concentrations, lime or other sources of alkalinity must be added. The chemical reaction between alum and lime alkalinity is as follows:

$$Al_2(SO_4)_3 \cdot 18H_2O + 3Ca(OH)_2 \longrightarrow 2Al(OH)_3 \downarrow + CaSO_4$$

$$+ 18H_2O \quad . \quad . \quad . \quad . \quad . \quad . \quad . \quad . \quad . \quad . \quad . \quad . \quad . \quad . \quad . \quad . \quad . \quad . (12.2)$$

Treatment with lime produces non-carbonate hardness.

Alum and Soda Ash

Soda ash (sodium carbonate) may also be used as the source of alkalinity.

$$Al_2(SO_4)_3 \cdot 18H_2O + 3Na_2CO_3 + 3H_2O \longrightarrow 2Al(OH)_3 \downarrow + 3Na_2SO_4$$

$$+ 3CO_2 + 18H_2O \quad . \quad . \quad . \quad . \quad . \quad . \quad . \quad . \quad . \quad . \quad . \quad . \quad . \quad . \quad . \quad . (12.3)$$

In this case hardness is not produced, but soda ash is more expensive than lime or most other chemicals useful as sources of alkalinity. A concentration of alum between 5 to 34 mg/ℓ may be required for successful coagulation with most pre-settled waters.

Ferrous Sulfate

The chemical reaction with ferrous sulfate (copperas) is as follows

$$FeSO_4 \cdot 7H_2O + Ca(OH)_2 \longrightarrow Fe(OH)_2 \downarrow + CaSO_4$$

$$+ 7H_2O \quad \ldots \ldots \ldots \ldots \ldots \ldots \ldots (12.4)$$

Ferrous hydroxide is soon oxidized by the dissolved oxygen in water, yielding:

$$4Fe(OH)_2 + O_2 + 2H_2O \longrightarrow 4Fe(OH)_3 \downarrow \ldots \ldots (12.5)$$

Flocs of ferrous hydroxide are dense and settle quickly. The use of copperas requires good control of the process. If an excess of lime is present, precipitation may occur in the distribution system due to the reaction with bicarbonate alkalinity. Copperas is best suited for turbid water having sufficient natural alkalinity. Ferric hydroxide is more insoluble than ferrous hydroxide in water having normal pH values. Hence, ferrous hydroxide is more useful as a coagulant when sufficient oxygen is present.

Ferric Coagulation

A mixture of ferric chloride ($FeCl_3$) and ferric sulfate ($Fe_2(SO_4)_3$), known as chlorinated copperas has been successfully used as a coagulant. This can be prepared by the addition of chlorine in ferrous sulfate and the following reaction takes place:

$$3(FeSO_4 \cdot 7H_2O) + 1.5Cl_2 \longrightarrow Fe_2(SO_4)_3 + FeCl_3$$

$$+ 21H_2O \quad \ldots \ldots \ldots \ldots \ldots \ldots (12.6)$$

Both ferric sulfate and ferric chloride react with the alkalinity

$$2Fe_2(SO_4)_3 + 6Ca(HCO_3)_2 \longrightarrow 4Fe(OH)_3 + 6CaSO_4$$

$$+ 12CO_2 \quad \ldots \ldots \ldots \ldots \ldots \ldots (12.7)$$

$$2FeCl_3 + 3Ca(HCO_3)_2 \longrightarrow 2Fe(OH)_3 + 3CaCl_2$$

$$+ 6CO_2 \quad \ldots \ldots \ldots \ldots \ldots \ldots (12.8)$$

With chlorinated copperas the effective range of pH values in which coagulation can take place is greater.

Ferric Chloride

Soft, highly colored waters are not effectively treated with alum. When ferric chloride is added to such a water, insoluble ferric hydroxide flocs are formed.

$$FeCl_3 + 3H_2O \longrightarrow Fe(OH)_3 + 3H^+ + 3Cl^-$$

Water containing hydrogen sulfide can be treated successfully with ferric chloride.

12.5.2 Limitations of Reaction Equations

The chemical reactions give an idea as to how the reactions proceed, but the equations cannot be used to predict stoichiometric relationships. Tests must be conducted to determine the actual amount of chemicals required.

12.5.3 Effect of pH

The control of pH is very important for the precipitation of colloids. The point of optimum flocculation depends upon the nature of the water as well as the pH value. In case of highly colored soft water, the point of optimum flocculation may be at a pH value of 3.8 and for other waters at pH values greater than 8. In general the optimum pH for flocculation lies in the range of 5 to 6.5. The pH value of water influences coagulation because solubility of chemical compounds depends upon it. The solubility constants for hydroxides of aluminum and iron, which are the common coagulants, are given in Table 12.1.

12.6 DETERMINATION OF COAGULANT DOSE IN THE LABORATORY

Reactions during coagulation are complex, and it is impossible to determine correctly the coagulant dose for a particular water with the simple chemical reaction equations shown in Section 12.5.1. Coagulation is influenced by factors such as pH value, temperature, alkalinity, turbidity, and agitation. Therefore, the jar test is conducted to determine the dose required for effective coagulation. The test is conducted with 1 liter beakers and a stirring device such as that shown in Figure 12.2. The following procedure for the jar test yields satisfactory results.

    1.   Determine the color, turbidity, pH value, and alkalinity of water sample.

Table 12.1. Solubility Constants for Hydroxides of Aluminum and Iron

Compound	Ion Product	K_s at 25°C	pH	Influence of pH on Solubility
$Al(OH)_3$	$(Al^{+++}) (OH^-)^3$	1.9×10^{-33}	4.0	Decreases with an increase
$Fe(OH)_3$	$(Fe^{+++}) (OH^-)^3$	4×10^{-38}	3.0	''
$Fe(OH)_2$	$(Fe^{++}) (OH^-)^3$	1.65×10^{-15}	8.0	''

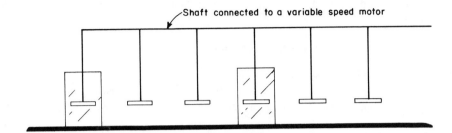

Figure 12.2. Stirring device for jar test.

2. Place a 1000 ml sample in each beaker, and then add the coagulant.
3. Mix the water and coagulant rapidly for 1 minute and then mix slowly for 15 to 20 minutes.
4. Note the time of appearance of a visible floc in each of the beakers.
5. Allow the mixture to settle 40 to 45 minutes.
6. Withdraw samples with a pipette without disturbing the settled matter.
7. Determine the color, turbidity, pH and alkalinity of the supernatant.
8. Select the optimum dose for maximum removal of color and turbidity.
9. The pH value may also be varied and the optimum value for effective coagulation determined by repeating steps 1 through 8.

12.7 COAGULANT AIDS

Certain chemicals improve the coagulation process by producing larger floc size, reducing the dosage of metal coagulant, and removing organic color from water. Activated silica is a short chain polymer which has binding properties, and when used with alum, a tougher, more settleable floc is produced. The required dose may be between 5 and 10 mg/l.

Polyelectrolytes are long chain polymers, which act as a polyvalent ion. Polymer chains have a very large number of ionic sites, and coagulation is promoted by neutralization of the negative charge on colloids by these ionized sites. Electrostatic cross connections also bind the linear chain together. The polyelectrolyte causes a reduction in the zeta-potential without an appreciable change in alkalinity or pH value.

Nirmali seed *(Strychnos potatorum, linn)* is a source of natural polyelectrolyte. Figure 12.3 shows the turbidity removals obtained with

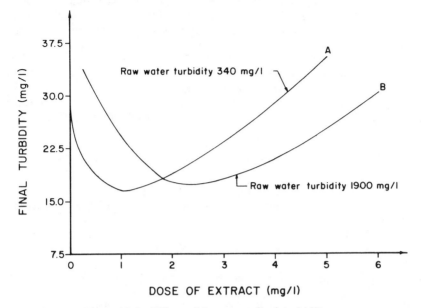

Figure 12.3. Effect of dosage on final turbidity.

various dosages of Nirmali seed extract. With doses as low as 1 to 2.5 mg/ℓ, a final turbidity between 15 and 20 mg/ℓ could be achieved. When used in conjunction with alum, coagulation is improved.

12.8 FACTORS INFLUENCING THE COAGULATION PROCESS

The following factors are known to influence the coagulation process:
1. Water quality.
2. The quantity and characteristics of colloidal matter.
3. pH value of the water.
4. Fast mixing, flocculation time, and the speed of paddles.
5. Temperature.
6. Alkalinity.
7. Characteristics of ions in water.

12.9 OPERATIONAL VARIABLES IN TURBIDITY REMOVAL

Experience and research have shown paddle speed, followed by flocculation time and alum dosage to be the most significant independent variables.
1. The optimum paddle speed ranges between 40 and 50 rpm.
2. The optimum flocculation time is approximately 30 minutes.

3. The significant interactions between the independent variables (alum dosage, flocculation time, and paddle speed) can be utilized in more efficient water treatment. For example a higher percentage removal of turbidity can be obtained by using low alum dosages, and increasing the flocculation time and/or paddle speed. On the other hand, high alum dosages and reduced flocculation times and/or paddle speeds can be utilized.

4. The above variables emphasize the importance of the following equipment and methods in the removal of turbidity in water treatment plants:
    a. Variable speed flocculator paddles.
    b. Multiple flocculation basins that can be used in series and parallel operation.
    c. The use of jar tests to determine the optimum alum dosage continuously during the operation of water treatment plants.

12.10 MECHANISMS OF COAGULATION

The mechanisms of coagulation are very complex and coagulation occurs as a combination of the following:
1. Coagulation of colloidal particles may occur through cross-linking at the bridging of segments of polymer molecules.
2. The adsorption of the hydrolysis products by the surface of the particle may result in the neutralization of the surface charges.
3. Simple enmeshment of the colloidal particles in the hydroxide flocs may result in settlement of the suspended particles.
4. The aluminum ions (or other ions) may act as electrolytes in the double layer model of the charged particles.

12.11 MIXING

Satisfactory coagulation requires that the chemicals and the raw water be thoroughly mixed. To achieve mixing, chemicals are added to the water in channels at hydraulic jumps, or at venturi meters, or just prior to entering pumps, or in mixing basins. Basins are usually fitted with impeller turbines or propellers moving at high speeds (Figure 12.4). The detention time in mechanically agitated tanks may be between 30 to 60 secs. The loss of head in such tanks is about 1 m.

12.11.1 Theory of Mixing

When chemicals are added to water, there should be thorough mixing so that uniform intermingling of water particles and chemicals takes place. For this purpose turbine impellers, or propellers, are commonly used (Figure 12.4). The power, P, required for mixing is a function of the diameter of the

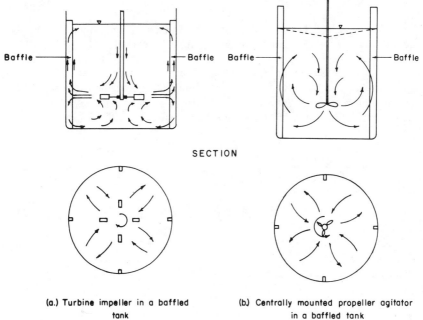

SECTION

(a.) Turbine impeller in a baffled
tank

(b.) Centrally mounted propeller agitator
in a baffled tank

PLAN

Figure 12.4. Mixing devices.

mixer, d, its speed, N, the absolute viscosity, μ, density of the fluid, D, and the acceleration due to gravity, g:

$$P = f(d, \mu, g, D, N) \quad \ldots \ldots \ldots \ldots \ldots \ldots \quad (12.9)$$

Using dimensional analysis techniques, we get:

$$\frac{P}{d^5 D N^3} = K \left(\frac{d^2 ND}{\mu}\right)^{-b} \left(\frac{dN^2}{g}\right)^{-c} \quad \ldots \ldots \ldots \ldots \quad (12.10)$$

Where K, b, and c are constants, let $\dfrac{P}{d^5 DN^3} = N_p$ = power number

The Reynold's Number, R_e, is VdD/μ, and assuming $V \propto Nd$,

$$R_e = \frac{Nd^2 D}{\mu}$$

The Froude Number, F_r, is $\dfrac{V^2}{gd} = \dfrac{N^2 d}{g}$

Therefore,

$$N_p = K(R_e)^{-b} (F_r)^{-c} \quad \ldots \ldots \ldots \ldots \quad (12.11)$$

The values of K and the exponents b and c will depend upon the mixing conditions. The relationship between N_p and R_e is shown in Figure 12.5. The curve 1-2-3-4 corresponds to agitation without vortex formation, and the curve 1-2-5 corresponds to agitation with vortex formation. For lower values of Reynold's Number ($R_e < 300$), both the curves coincide with each other and remain a straight line up to $R_e = 10$. In such a case the value of the exponent c will be zero and b will be approximately one. Therefore,

$$N_p = \left(\frac{Nd^2 D}{\mu}\right)^{-1} \quad \ldots \ldots \ldots \ldots \ldots \quad (12.12)$$

and

$$P = K \left(\frac{Nd^2 D}{\mu}\right)^{-1} d^5 N^3 D = K (\mu N^2 d^3) \quad \ldots \ldots \quad (12.13)$$

The portion 3-4 of the curve 1-2-3-4 represents the fully developed turbulent condition without vortex formation. This becomes horizontal after $R_e = 10,000$, and the value of b in this range will be zero.   Hence

$$N_p = K \quad \ldots \ldots \ldots \ldots \ldots \ldots \ldots \quad (12.14)$$

and

$$P = KDN^3 d^5 \quad \ldots \ldots \ldots \ldots \ldots \quad (12.15)$$

The portion of curve 1-2-5 in the turbulent region is irregular indicating that the Froude Number is effective when there is turbulence accompanied by vortex formation. An equation has not been developed for this portion of the curve.

The values of K for various shapes of impellers are given in Table 12.2. These values correspond to the impellers rotating at the center line of a flat bottom cylindrical tank. Four baffles are located on the tank walls, and the width of baffles equals 10 percent of the tank diameter. The liquid depth is equal to one tank diameter. The impeller diameter is one third of the tank diameter and the impeller is placed at one diameter above the tank bottom. To use the above expressions and values of K in Table 12.2, it is necessary that the conditions of use be essentially the same as the experimental conditions. Otherwise it is necessary to determine other values of K.

12.12 FLOCCULATION

By stirring and agitating the liquid, chances of contact between the small flocs is increased and this is called flocculation. Generally paddle type mixers are used in a water treatment plant for flocculation (Figures 12.6 and 12.7).

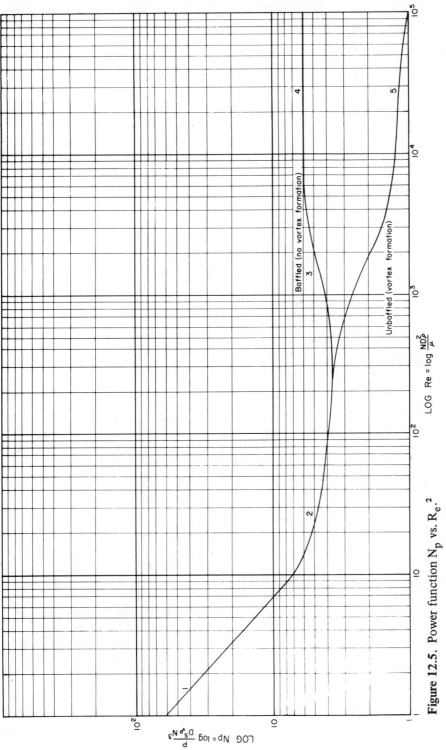

Figure 12.5. Power function N_p vs. R_e.

LOG R_e = log $\dfrac{ND^2\rho}{\mu}$

LOG N_p = log $\dfrac{P}{D_s^5 \rho N^3}$

Baffled (no vortex formation)

Unbaffled (vortex formation)

Table 12.2. Values of Constant K^2

Impeller[a]	Viscous Range	Turbulent Range
Propeller, square pitch, three blades	41.0	0.32
Propeller, pitch of two, three blades	43.5	1.000
Turbine, six flat blades	71.0	6.30
Turbine, six curved blades	70.0	4.80
Fan turbine, six blades	70.0	1.65
Flat paddle, two blades	36.5	1.70
Shrouded turbine, six curved blades	97.5	1.08
Shrouded turbine with stator (no baffles)	127.5	1.12

[a]A helix is traced by a revolving propeller in a fluid. Depending upon the inclination of the propeller blades, the fluid will be displaced longitudinally a fixed distance in every revolution of the propeller. The ratio between this distance and the diameter of the propeller is called the pitch. The propeller has a square pitch when this ratio is 1.

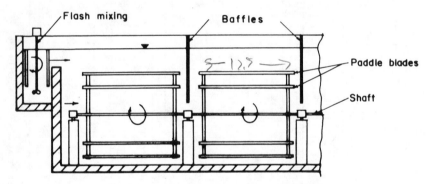

Figure 12.6. Flocculating basin. (Courtesy of TH-Dresden.)

To develop a relationship for describing flocculation assume that particles of diameter d_1 and d_2 are settling in a viscous fluid. Assume that the numbers of d_1 and d_2 particles are n_1 and n_2, respectively.

These particles can come in contact when their centers are at a distance of $\frac{1}{2}(d_1 + d_2)$ (Figure 12.8). Therefore, n_2 number of particles having a diameter d_2 per unit volume, dv, flowing through the sphere of radius $\frac{1}{2}(d_1 + d_2)$ will come into contact with particles of diameter d_1 per unit of time. The flow through a thin lamina dx per unit of time will be

$$dQ = \left(x \frac{dv}{dy}\right) 2 (r^2 - x^2)^{\frac{1}{2}} dx \quad . . . . (12.16)$$

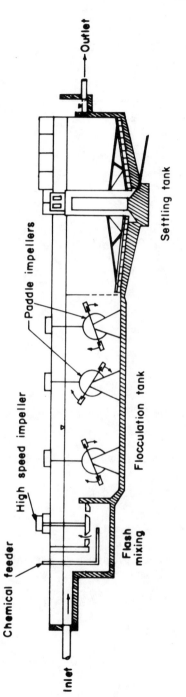

Figure 12.7. Clarifier (Dorr mechanically operated mixing flocculation and settling tank). (Courtesy of Dorr-Oliver, Inc.)

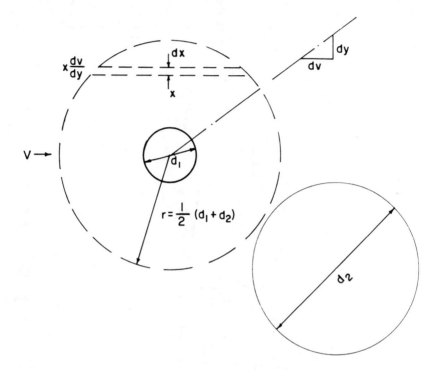

Figure 12.8. Flocculation of particles.[3]

Integrating will yield the total flow through the sphere of radius ½ $(d_1 + d_2)$

$$Q = 2 \int_o^r \frac{dv}{dy} (d_1 + d_2)^3$$

$$= (1/6) \frac{dv}{dy} (d_1 + d_2)^3 . \quad . \quad . \quad . \quad . \quad . \quad . \quad . \quad . \quad . \quad . \quad (12.17)$$

The total number of contacts per unit time, per unit volume between n_1 and n_2 number of particles of diameter d_1 and d_2 will be,

$$N' = (1/6)n_1 \cdot n_2 \frac{dv}{dy} (d_1 + d_2)^3$$

and for the whole system, it will be

$$N = (1/6) n_1 \cdot n_2 \frac{dv}{dy} (d_1 + d_2)^3$$

$$= (1/6) n_1 \cdot n_2 G (d_1 + d_2)^3 \quad . \quad . \quad . \quad . \quad . \quad . \quad . \quad (12.18)$$

in which dv/dy = G = temporal mean velocity gradient. From Equation 12.18 it follows that the number of contacts is directly proportional to the mean velocity gradient of the system and that the larger the concentration and the bigger the size of particles, the better will be the flocculation.

With an increase in velocity gradient, the shear stress increases ($\tau = \mu$ (dv/dy)). The maximum size of the floc is dependent on the velocity gradient, because at higher values of velocity gradient the floc will break. Therefore, the maximum size of the floc to be settled will limit the velocity gradient.

12.13 RELATIONSHIP BETWEEN POWER CONSUMPTION AND VELOCITY GRADIENT

Considering the forces on an element in a flowing viscous fluid (Figure 12.9),

$$(p\, \partial y\, \partial z) + \left(\tau + \frac{d\tau}{dy}\, \partial y\right)\, \partial x\, \partial z = \left(p + \frac{dp}{dx}\, \partial x\right)\, \partial y\, \partial z$$

$$+ \tau\, \partial x\, \partial z$$

$$\frac{d\tau}{dy}\, \partial y\, \partial x\, \partial z = \frac{dp}{dx}\, \partial x\, \partial y\, \partial z$$

$$\frac{d\tau}{dy} = \frac{dp}{dx} \quad \cdots \cdots \cdots \cdots \cdots \cdots \quad (12.19)$$

Where p = intensity of pressure and τ = shear stress.   The power required is the torque multiplied by the angular velocity

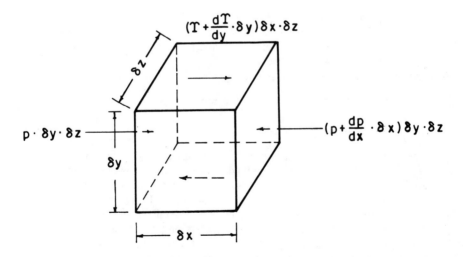

Figure 12.9. Forces on an element in a flowing viscous fluid.

$$P_1 = \text{Work done x angular velocity}$$

$$= (\tau \, \partial x \, \partial z) \, \partial y \, \frac{dv}{dy} \ldots \ldots \ldots \ldots \ldots \quad (12.20)$$

Therefore, power required per unit volume is

$$P = \frac{P_1}{\partial x \, \partial y \, \partial z} = \tau \frac{dv}{dy} = \mu \left(\frac{dv}{dy}\right)^2 = \mu G^2 \ldots \ldots \quad (12.21)$$

$$\therefore G = \sqrt{P/\mu}$$

Substituting the value of G in Equation 12.18, we get

$$N = \frac{1}{6} n_1 \, n_2 \, (P/\mu)^{\frac{1}{2}} \, (d_1 + d_2)^3 \ldots \ldots \ldots \ldots \quad (12.22)$$

According to Equation 12.22, the number of contacts increases with the power input. The flocculation and agglomeration of particles will increase with the number and size of particles, but there is a limiting size floc beyond which the shearing stress will break the flocs.

12.13.1 Paddle Blades (Figures 12.6 and 12.7)

In a flocculation device, paddle blades gently revolve either on a horizontal or vertical shaft agitating dilute suspensions. In such systems the power input is a function of the drag force of the paddle, which is expressed as:

$$F_d = \frac{1}{2} C_d A D v^2 \ldots \ldots \ldots \ldots \ldots \ldots \ldots \ldots \quad (12.23)$$

in which

C_d = coefficient of drag
A = area of the paddles (m^3)
v = velocity of the paddle relative to the water (m/sec)
D = density of water (kg/m^3)

Therefore P, the power input per unit volume, will be

$$P = \text{force x velocity} = C_d \frac{A D v^3}{2V} = \mu G^2$$

$$G = (C_d A D v^3 / 2 \mu V)^{\frac{1}{2}} \ldots \ldots \ldots \ldots \ldots \quad (12.24)$$

Where paddle impellers are used, stator blades as baffles may also be provided. In the absence of stator blades, the area of the paddle blades should not be more than 15 to 20 percent of the cross-sectional area of the tank. The peripheral speed of the paddles should be between 0.15 m/sec and 1 m/sec.

According to Camp,[3] the value of G generally varies between 35 and 66 m/sec-m in plants with paddle impellers and between 20 and 74 for baffled mixing basins. The product of G and detention time is between 10^4 and 10^5. It has been found that G values should not be less than 10 nor greater than 75.

12.13.2 Baffled Basins

When water is flowing at a rate, Q (m³/sec), through a basin of length, ℓ(m), and cross-sectional area, A(m²), as shown in Figure 12.10, and a head loss of h(m) is assumed, the power per unit volume of the fluid is:

$$P = \frac{DgQh}{A\ell} = \frac{vDgh}{\ell} = \frac{Dgh}{t} \quad \ldots \ldots \ldots \ldots \ldots \quad (12.25)$$

in which D is the density of the fluid, v is the velocity of flow, and t is the detention time.

$$\frac{Dgh}{t} = \mu G^2 \,;$$

hence

$$G = \left(\frac{Dgh}{\mu t}\right)^{\frac{1}{2}} \quad \ldots \ldots \ldots \ldots \ldots \ldots \ldots \quad (12.26)$$

The velocity of flow generally varies between 0.15 m/sec and 1 m/sec. The detention time may be between 10 and 90 min with a head loss of about 0.3 to 1 m.

Example 12.1

A flocculation tank is fitted with paddle impellers 6 m long mounted on two horizontal shafts perpendicular to the flow and rotating at a speed of 4 rpm. Each shaft is fitted with two paddles 20 cm wide fitted opposite each other. The center of the paddle is 2 m from the center of the shaft. The rate of flow of water is 10.5 m³/min and the detention time is 40 min. The coefficient of drag is 1.4. The mean velocity of the water is one fourth the paddle velocity. Calculate (1) the ratio of the paddle area and the cross sectional area of the tank expressed as percentage; (2) the velocity differential

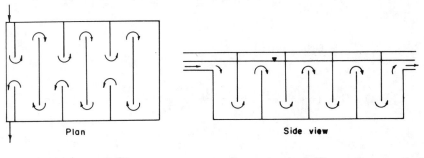

Plan	Side view
Round the end baffles	Over and under baffles

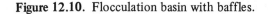

Figure 12.10. Flocculation basin with baffles.

for the paddles; (3) the power and the energy requirement; and (4) the value of G and Gt.

Volume of the tank = 10.5 x 40 = 420 m^3.

Let the depth = 5 m. Area = 420/5 = 84 m^2.

Assume length = 12 m, width = 7 m.

1. Paddle area rotating in the cross section = 6 x 2 x 0.2 = 2.4 m^2.

    Cross sectional area = 35 m^2.

    Percentage area = 2.4/35 x 100 = 6.85.

2. The linear velocity of paddle blades = (4)(4)π /60 = 0.837 m/sec.

    ∴ Velocity differential for the paddle = 0.837 x 0.75 = 0.63 m/sec.

3. Total area of the paddle = 3 x 2 x 2 x 2 x 0.2 = 4.8 m^2.

$$\text{Total power input } P_t = \frac{1}{2} C_d ADv^3$$

$$= \frac{1.4 \times 4.8 \times 1000 \times (0.63)^3}{2}$$

$$= 0.84 \text{ Kw}^+ (1.14 \text{ mhp})$$

$$\text{Total power input}/10000 \text{ m}^3 = \frac{0.84^2}{10.5 \times 60} \times 10000$$

$$= 1.331 \text{ Kwh}$$

$$G = \sqrt{\frac{C_d ADv^3}{2\mu V}} = \sqrt{\frac{1.4 \times 4.8 \times (0.63)^3 \times 100^2}{2 \times 1.004 \times 10^{-2} \times 420}}$$

$$= 45 \text{ m/sec-m}$$

∴ Gt = 45 x 40 x 60 = 10.80 x 10^4

Example 12.2

In a baffled basin the rate of flow of water is 12 m^3/min. The detention time is 40 min and the head loss is 1 m. Calculate (1) the power input, (2) the value of G and Gt.

$$(1) \quad P = \frac{Dgh}{t} = \frac{1000 \times 9.81 \times 1}{40 \times 60} = 4.1 \text{ Watt/m}^3$$

$$(2) \quad G = \left(\frac{Dgh}{\mu t}\right)^{1/2} = \left(\frac{1000 \times 9.81 \times 1}{1.004 \times 10^{-2} \times 40 \times 60}\right)^{1/2} = 20 \text{ m/sec-m}$$

$$Gt = 20 \times 40 \times 60 = 4.8 \times 10^4$$

$^+$1 KW = 1.36 mhp.

12.14 COMBINED FLOCCULATION AND CLARIFYING UNIT

Frequently a combined flocculation and settling tank, called a clariflocculator, is used (Figure 12.11). The depth of these tanks may be 3 to 7 m and the clarifying zone in the clariflocculator is approximately 2.5 m.

The detention time is generally from 2 to 6 hours, but it may be as short as 40 min in a clariflocculator. A continuous sludge removing mechanism is provided in the clarifying units.

Example 12.3

Design a flash mixer and a clariflocculator for a plant of 4550 ℓ/min (1200 gpm).

Solution:

Flash Mixer: Assume the detention time = 2 minutes.
    Volume of the flash mixer = $(2 \times 4550)/1000 = 9.1$ m^3 (321 ft^3).
    Assume the depth = 1.5 m.
    Hence area = $9.1/1.5 = 6.07$ m^2 (65.3 ft^2).
    Use dimensions of 2.46 x 2.45 x 1.5 m (8.08 x 8.08 x 5 ft).
Clariflocculator:
(a)   Flocculator
    Assume detention time = 30 minutes.
    Volume = $(4550 \times 30)/1000 = 136.5$ m^3 (4813 ft^3).
    Assume the depth = 3 m (10 ft).
    Area = $136.5/3 = 45.5$ m^2 (481.3 ft^2).
    Diameter = 7.62 m (24.76 ft).

(b)   Sedimentation tank
    Assume detention time = 4 hrs.
    Volume = $(4550 \times 4 \times 60)/1000 = 1092$ m^3 (38,503 ft^3).
    Use depth of 3.66 m (12 ft).
    Depth of sedimentation tank is 0.6 to 1.2 m more than the depth of the flocculating tank.
    Volume of the flocculator for the new depth = $\pi/4 \times 7.62^2 \times 3 = 136.74$ m^3 (4841 ft^3).
    Total volume of flocculator and sedimentation tank = $1092 + 136.74 = 1228.74$ m^3 (38,503 + 4841 = 43,344 ft^3).
    New surface area = $1228.74/3.66 = 335.72$ m^2 (3612 ft^2).
    Outside diameter = 20.68 m (68 ft).

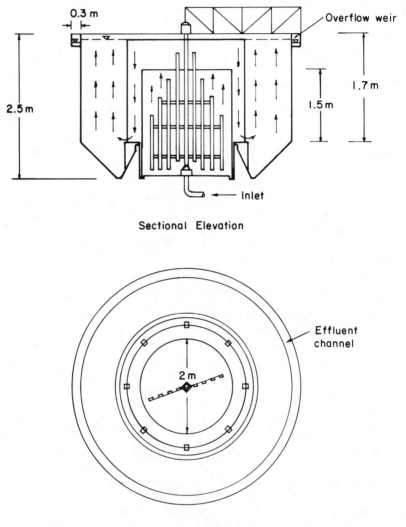

Sectional Elevation

Plan (without bridge)

Figure 12.11. Clariflocculator. (Courtesy of Dorr-Oliver, Inc.)

13. Filtration

13.1 GENERAL

Suspended and colloidal particles in water are not entirely removed by the sedimentation process. Water containing suspended and colloidal impurities, may be purified by filtration, a process which allows the water to pass through a bed of sand or a combination of granular materials. Natural filtration occurs when surface water percolates through the soil and joins the groundwater source. Natural filtration is not economical for water treatment.

Bacteria are effectively removed by filtration. Filtration also helps in the removal of color, taste, odor, iron, and manganese. A slow sand filter was first used in the United Kingdom early in the Nineteenth Century, and a rapid sand filter was developed in the United States between 1900 – 1910.[1] Modifications of the rapid sand filter and other types of filters are continuously being developed, i.e. pressure sand filters, multiple media filters, diatomateous earth filters, upflow filters, etc.

13.2 THEORY OF FILTRATION

When water containing suspended impurities passes through a bed of sand, many of these impurities are removed. The removal of these substances in the pores of sand occurs due to a combination of physical and chemical processes. Straining phenomenon and the adsorption of particles due to opposite electric charge influences the filtration process. The exact process by which filters remove suspended matter is still not clear; however, some of the factors which contribute to the removal are:

1. Straining action taking place at the surface of the filter bed.
2. Sedimentation inside the filter bed.
3. Contact of the floc particles with the surface of grains or with already deposited flocs. (This takes place due to the convergence of stream lines at contractions in the pore channels and near the curved surfaces of the grains.)
4. Adsorption or electrokinetic forces.
5. Coagulation inside the filter bed.
6. Biological activity depending upon the concentration of the organic matter present in water.
7. Colloidal mesh structure of filtering media.

Suspended particles too large to pass through the pores of the sand bed will be retained within the bed. Colloidal matter and bacteria, small in size, may not be completely removed. The space between the sand grains acts as a sedimentation tank in which the suspended matter settles upon the sand particles. Colloidal matter may be arrested due to adsorption or electrokinetic forces. Much of the dissolved and colloidal matter not converted to flocs in the flocculation and settling tanks coagulates inside the filter and is removed.

Substances retained during filtration are collected on the surface of the filter medium and form a mesh structure which functions as a porous layer and helps in the cleaning action. If sufficient nutrient is available, organisms begin to grow on the surface of the filter. As a result a mat is formed containing slimy "Zoogleal" organisms known as "Schmutzdecke" (dirty cover). This mat helps in the straining action of the filter, but must be removed when the head loss through the filter is high. Such a mat is undesirable in a rapid sand filter because it encourages formation of mud balls during backwashing.

Filter performance formulations describing the performance of the system as a whole and including both transport and attachment have been attempted. Iwasaki[2] developed a simple rate equation and a material balance expression in 1937:

$$\frac{\partial C}{\partial \ell} = -\lambda C \qquad \qquad \qquad (13.1)$$

$$\frac{\partial \sigma}{\partial t} = -\frac{v_s}{(1 - f_\sigma)} \frac{\partial C}{\partial \ell} \qquad \qquad (13.2)$$

in which
$$
\begin{aligned}
C &= \text{suspension concentration in volumetric units}\\
\ell &= \text{depth of the media}\\
\lambda &= \text{filter efficiency constant}\\
\sigma &= \text{specific deposit, volume of deposit per unit volume of filter}\\
\frac{\partial \sigma}{\partial t} &= \text{rate of buildup of the deposit}\\
\frac{\partial C}{\partial \ell} &= \text{rate of removal of suspension}\\
v_s &= \text{approach velocity}\\
f_\sigma &= \text{deposit porosity}
\end{aligned}
$$

The first order reaction equation (Equation 13.1) describes the performance for a system with uniform media and a homogeneous suspension. The probability that a particle will reach the media surface in traveling a distance $\partial \ell$ is proportional to the number of particles present. Iwasaki recognized that λ is not constant and developed Equation 13.2 to relate buildup with time. Because of the variability of the density of the suspended particles and the deposit porosity and the fact that λ is not constant with time and space, the above expressions are of little practical value. Ives et al.[3,4] and Fox and Cleasby[5] have modified the above expressions and performed experiments with carefully controlled media and particles, and again it appears that these types of formulations have little practical value.

Head loss during filtration and the length of filter run have been described by expressions incorporating σ and f_σ, and as with the performance relationships, the variation with time and space render the expressions essentially useless in terms of making predictions based on the original characteristics of the media.

The ideal filtration system is one in which the suspension is trapped on the media in such a manner that limiting head loss and suspension of effluent concentration are attained simultaneously. This is achieved best when the flow is from coarse to fine media (Figure 13.1). This system differs considerably from traditional practice, but it is illustrated by the theoretical expressions which, however, do not adequately define the process. By maintaining a constant hydraulic gradient through the depth of the bed, the overall head loss is minimized and higher flow rates are possible. With the finer material at the bottom of the filter, removal of the small residual turbidity would be expected. This filtration scheme is frequently referred to

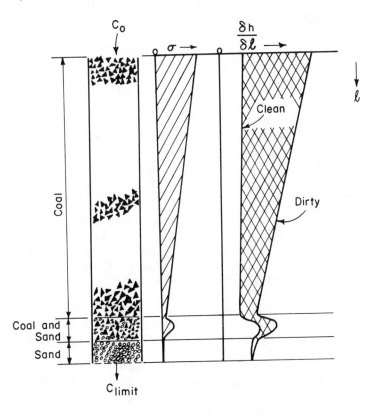

Figure 13.1. Balanced filter operation.

as "in-depth" filtration. The characteristics of the suspension and conditioning with polyelectrolytes may also influence filter performance.

Rapid sand filters are commonly employed in municipal water treatment plants, and in some areas slow sand filters are used. These two types along with modifications and other type filters are discussed in the following sections.

13.3 SLOW SAND FILTER

The natural filtration which takes place during percolation of water into the ground is very similar to a slow sand filter. In slow sand filtration water is passed through a bed of sand after long periods of settling without any chemical pretreatment. During operation the head of water above the bed of sand is between 0.9 to 1.6 m. The velocity of water (rate of filtration) in the filter is between 0.1 and 0.4 m/hr (0.1 to 0.4 m^3/m^2-hr). The effective size of sand particles may be between 0.25 and 0.35 mm (i.e. the sieve size in millimeters that permits 10 percent of sand by weight to pass) and the uniformity coefficient between 2 and 3 (i.e. the ratio between the sieve size that will pass 60 percent and the effective size).

Thickness of the sand bed may be 1.0 to 1.5 m supported on a layer of gravel 0.3 to 0.5 m thick (Figure 13.2). Gravel is placed in five or six layers with the finest size at the top (Table 13.1). The supporting layer of gravel prevents the penetration of fine sand particles into the lower layers and the removal of fine particles along with the filtered water. Treated water is collected in open jointed laterals placed beneath the gravel at 1.0 to 3.0 m centers. The diameter of the laterals may be between 100 and 300 mm. Perforated plastic pipe can also serve as the collection pipe.

A high degree of treatment is achieved with a slow sand filter. Slow sand filters are sensitive to large variations in the pH value of inflow. The formation of the "Schmutzdecke," which depends upon the growth of microorganisms, requires time to build up and is dependent upon the temperature. When a sudden gush of turbid water is applied to the filter bed, the pores of the film on the top surface become plugged resulting in the reduction in length of filter run. Because of this plugging of the pores, large settling tanks always precede a slow sand filter. Raw water must be given sufficient time to settle, sometimes weeks, before it is applied to the filter bed.

In some locations fine sieves, or screens (microstrainers), are used instead of large settling tanks. This method is popular in England, especially for the removal of plankton, which multiply rapidly in settling tanks where water may remain for a week.

The greatest disadvantage of a slow sand filter is that a large area is required for the filter beds as well as for the settling tanks. Since the sand bed cannot be backwashed, the operation of the filter must be stopped when head loss increases to about 1 meter so that the bed can be cleaned. The top surface of the filter bed requires cleaning either by hand or by mechanical

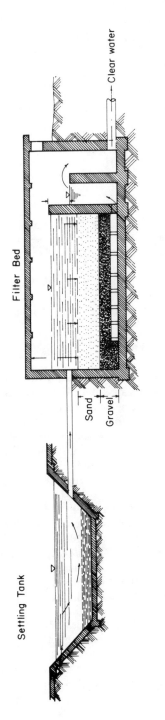

Figure 13.2. Slow sand filter with settling tank.

Table 13.1. Grading and Layer Thickness of Underdrain System For a Slow
Sand Filter

Size (mm)	Depth (cm)	Material
2 - 7	3 - 5	Coarse Sand
8 - 15	5 - 10	Gravel
15 - 30	10 - 15	Gravel
50 - 100	10	Packed Stone
100 - 200	Underneath	Not defined

means. A few centimeters (0.5 to 2.5 cm) of sand from the top is removed and replaced with clean sand. It may take several days before the filter bed can again be effective in treating water, because it takes some time for the mat to build up.

Slow sand filters are unsuited for treating water containing large quantities of organic matter. They can be used for small communities where the raw water quality is better than those of urban and industrialized areas.

13.4 RAPID SAND FILTER

In water purification the rapid sand filter has achieved much greater importance than the slow sand filter. The rapid sand filter was developed in 1860 in the U.S.A. and has become accepted in other countries since the first world war. The velocity of flow in a rapid sand filter is very high and the rate of filtration may be between 4 and 5 m^3/m^2-hr (sometimes the rate of filtration may be as high as 6 m^3/m^2-hr). The effective size of sand generally used in the bed is 0.45 - 0.55 mm. The uniformity coefficient of sand particles does not exceed 1.5. The bed is stratified due to backwash, the finer being at the top and the heavier at the bottom. The sand bed comprises 38 - 60 cm of gravel at the bottom and 80 cm of sand at the top (Figure 13.3). A 7.5 cm layer of coarse sand having an effective size between 0.8 and 2.0 mm is occasionally placed between the filter sand and gravel. The coarse sand should be free from impurities such as clay, dust, etc. The size and thickness of the corresponding layers of gravel are given in Table 13.2.

The level of water above the top of the bed may be between 0.9 and 1.6 m. The loss of head in a clean bed may be 0.2 to 0.5 m. The designed loss of head during operation may be about 3 m with a free board of about 1 m. Raw water is treated with chemicals for the removal of turbidity before entering the filter bed. It passes through the flocculation and settling tank, and then to the filter bed.

The process of cleaning and removal of suspended and colloidal matter in a rapid sand filter is different from that of the slow sand filter. Backwashing, or a reversal of the direction of flow is performed as soon as the

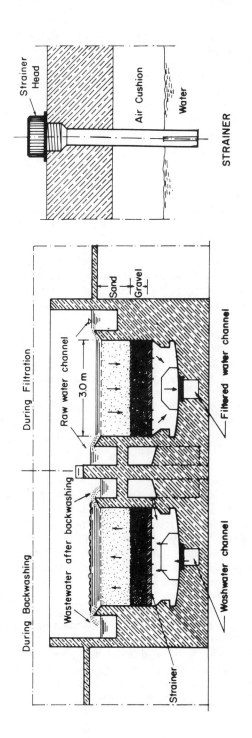

Figure 13.3. Rapid sand filter. (Courtesy of TH-Dresden.)

Table 13.2. Grading and Layer Thickness of Underdrain System For a Rapid
Sand Filter

Size in mm		Thickness in cm
65 - 38		13 - 20
38 - 20		8 - 13
20 - 12		8 - 13
12 - 5		5 - 8
5 - 2		5 - 8
	Total Thickness	39 - 62

designed loss of head is attained. A relatively complicated underdrain system
is required in a rapid sand filter to collect the filtered water and evenly
distribute the backwash water throughout the filter bed. Design criteria for
underdrains and filters in general are based essentially on experience. The
total area of openings in the underdrain laterals of a perforated pipe system
should be 0.3 to 0.5 percent of the filter surface area. The total cross
sectional area of the laterals should be about three times that of the strainer
or the lateral openings. The cross sectional area of the manifold may be 1.5 to
2 times the total area of the laterals. Perforated pipe under drains may be
made of brass, cast iron, or asbestos cement. Laterals are attached to the
manifold which carries the total flow of the filtered water.

A false bottom made of concrete, fitted with porcelain, plastic, or
metal strainers placed at 20 cm centers can also be used to simplify the
distribution of wash water. A filter bottom fitted with porous plates and
fastened to supports made of concrete may also be used, but these are easily
clogged. (See Figure 13.4 for types of underdrain systems employed.) When
compressed air is used in addition to wash water for backwashing the filter,
special types of strainers must be fitted to the filter bottom (Figure 13.3). To
distribute the air-water mixture uniformly in the entire bed, the air cushion
below the filter bottom is forced along with the wash water. Compressed air
(5 m head of water) is forced inside the air cushion pushing a mixture of
air-water inside the strainer through the slit. About 1 to 1.5 m^3 of free air per
square meter per minute is forced through the underdrains. The wash water
tank and the piping for the backwash system are designed based on 50
percent sand expansion.

Wash water systems may be designed for rates up to 56 m/hr. The
quantity of water in the wash water tank should be sufficient to wash one
filter for at least 10 minutes at the designed rate of backwashing. In general
the quantity of wash water used is between 1 and 6 percent of the filtered
water. Many water treatment plants have surface wash in addition to the back
wash system. A grid of horizontal pipes is provided at the top of the filter
bed, and vertical pipes are attached at 60 - 75 cm centers. These vertical pipes

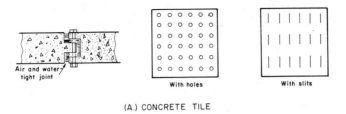

(A.) CONCRETE TILE

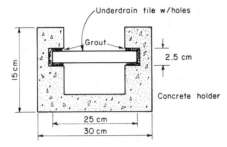

(B.) VITRIFIED CLAY TILE

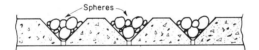

(C.) WHEELER UNDERDRAIN

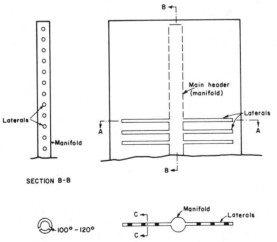

(D) PERFORATED PIPE

Figure 13.4. Types of underdrains.

are approximately 10 cm above the top of the sand bed and their bottom ends are perforated with 2.5 mm holes. The water is applied under a pressure of 1 to 2 atmosphere. The size of a large unit may be 5.4 x 6 m. Smaller units may vary between 2.7 x 3 m and 3.6 x 4.5 m. Hydraulically powered rotating perforated pipes are also used above the sand surface. A flow diagram for a rapid sand filtration plant is given in Figure 13.5.

Several proprietary rapid sand filter systems are available and provide features which take advantage of multiple media filtration and automatic continuous backwash. Figure 13.6 shows an automatic gravity filter with a self-contained backwashing system containing anthracite coal and sand media. The unit is adaptable and can be installed in multiple units to meet capacity requirements. Most equipment manufacturers produce a similar filter, but there may be many combinations of media and control devices. An automatic backwash filter which continuously backwashes a portion of the filter media is shown in Figure 13.7.

A detailed engineering analysis should be made before deciding on the type system to be installed. If proprietary devices offer operational or economic advantages, their use should be given careful consideration. The performance of these proprietary units has been established from experience and experimentation and the recommendations of the manufacturers must be followed.

13.5 PRESSURE FILTERS

Pressure filters are identical to gravity filters except that the unit is housed in an airtight steel chamber designed to handle pressures up to 10 atmospheres (Figure 13.8). The rate of filtration may be as high as 15 m/hr, but such a high rate should be used with caution. Even a filtration rate of 4.8 m/hr should be adopted with care.

The bacterial quality of the effluent from these filters is poor if the filter is not properly designed and operated. Pressure filters may be used in treatment of water for general use, e.g. washing, cooling, and industrial processes. Pressure filters are useful as a pretreatment unit for highly turbid raw waters. The effluent from this is then applied to a rapid or slow sand filter, or to another moderately loaded pressure filter.

In areas where the water is supplied under pressure, an advantage of using pressure filters is that the water, after passing through the filter, does not require further pumping.

13.6 DIATOMITE FILTERS

Diatomite filtration has been used successfully to treat domestic water supplies but has been used principally for swimming pool water treatment, industrial applications, and for emergency military and civil defense water supply production. There are instances where it can be competitive in small systems and should not be eliminated from an evaluation *a priori*.

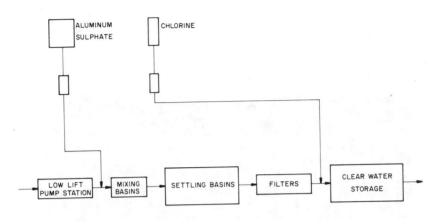

Figure 13.5. Flow diagram for rapid sand filter plant.

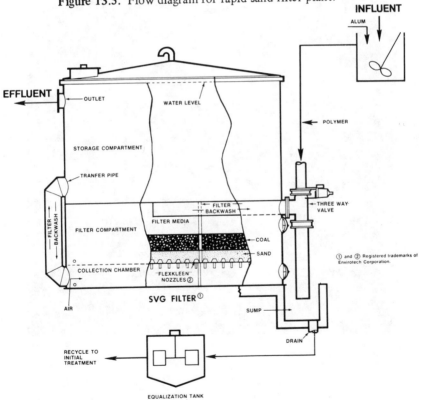

Figure 13.6. Typical flow sheet for solids removal by granular media filtration. (Courtesy of Envirotech Corp.)

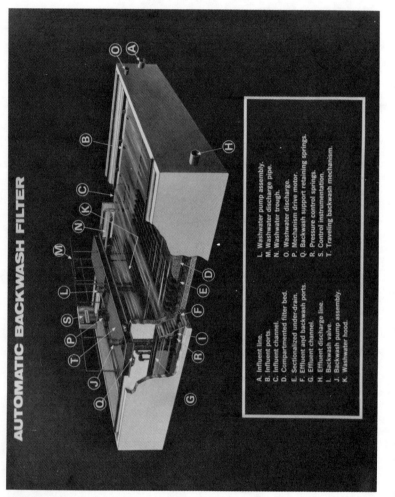

Figure 13.7. Automatic backwash filter. (Courtesy of Environmental Elements Corporation.)

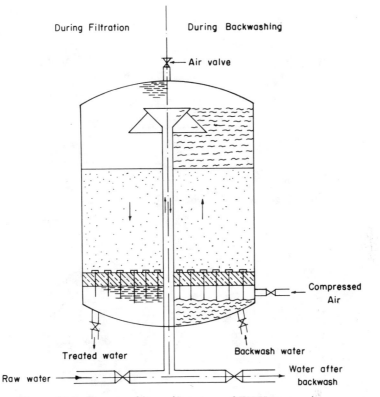

During Filtration   During Backwashing

Air valve

Compressed
Air

Treated water   Backwash water

Raw water   Water after
backwash

Figure 13.8. Pressure filter. (Courtesy of TH-Hannover.)

Diatomite is the fossil-like skeletons of diatoms ranging in size from less than 5 to over 100 microns (Figure 13.9). Many deposits of diatoms occurred in the geological past and are mined and processed to make a filter media. The skeleton is basically silica and is very porous.

Filtration with diatomite is performed by attaching the diatomite to the surface of a septum usually contained in a pressurized container (Figure 13.10). The initial application of diatomite is referred to as the "precoat." Water containing the suspension to be removed is then applied under pressure and forced through the diatomite coat trapping the particles. To prolong filter runs diatomite is added continuously with the water being treated to keep the coat porous. This is called "body feed." The precoat and body feed steps are illustrated in Figures 13.11 and 13.12, respectively. The basic components of a diatomite filtration system are shown in Figure 13.13. The systems can also be operated as vacuum filters.

Septum materials can be made of woven synthetic cloth, wire cloth, sintered grains of aluminum oxide, stainless steel, bronze or other non-woven media such as spaced wire helixes and various types of felts.

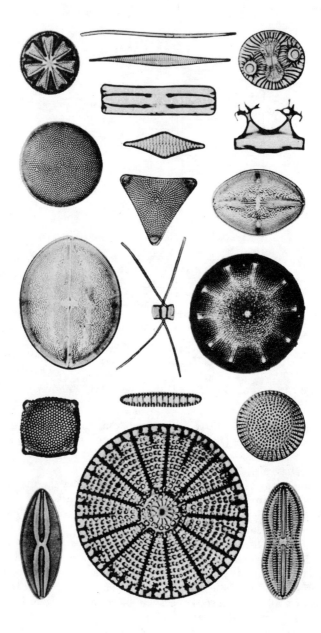

Figure 13.9. Photomicrograph showing diatom shells found in diatomite deposits. (Courtesy of Johns-Manville Co.)

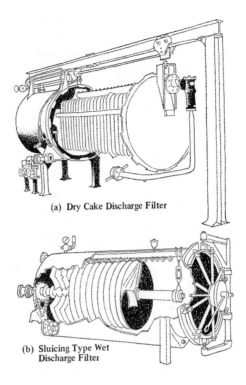

(a) Dry Cake Discharge Filter

(b) Sluicing Type Wet
Discharge Filter

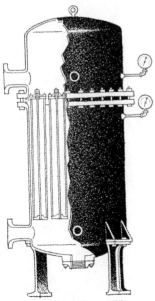

Figure 13.10. Type of pressure filters
used in diatomite filtration.
(Courtesy of Johns-Manville Co.)

(c) Cylindrical Element Filter

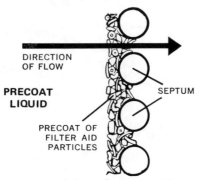

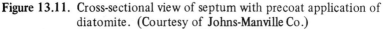

Figure 13.11. Cross-sectional view of septum with precoat application of diatomite. (Courtesy of Johns-Manville Co.)

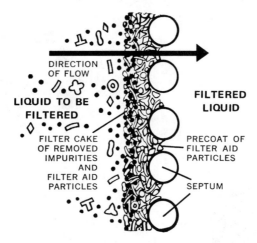

Figure 13.12. Cross-sectional view of septum during operation. (Courtesy of Johns-Manville Co.)

13.7 HYDRAULICS OF FILTRATION

13.7.1 Carman-Kozeny Equation

According to the Darcy-Weisbach equation

$$h_f = f\frac{\ell v^2}{2gd} \quad . \quad . \quad . \quad . \quad . \quad . \quad . \quad . \quad . \quad . \quad . \quad . \quad (13.3)$$

This equation may be used for channels in beds if d is replaced by hydraulic mean radius;

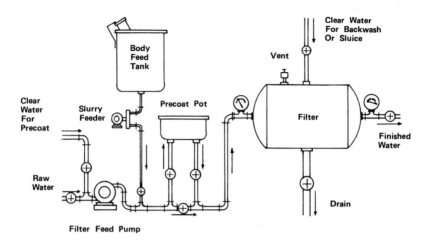

Figure 13.13. Basic components of a diatomite filtration system. (Courtesy of **Johns-Manville Co.**)

$$R = \frac{d}{4} \quad \text{or} \quad d = 4R$$

Hence
$$h_f = f\frac{\ell v^2}{8Rg} \quad \ldots \ldots \ldots \ldots \ldots \ldots \quad (13.4)$$

in which

ℓ = depth of bed over which head loss occurs
f = friction coefficient (dimensionless)
R = hydraulic mean radius
v = average velocity of fluid through the channels

Total channel volume = total volume of voids = V

$$e = \frac{V_u}{V}$$

$$V_{s_1}e = V_{s_1}\frac{V_u}{V} = V_u(1-e)$$

$$\therefore V_u = \frac{e}{1-e}V_{s_1} \quad \ldots \ldots \ldots \ldots \ldots \quad (13.5)$$

in which

e = porosity of bed
V_{s_1} = volume of solids = NV_p in which N = number of particles and V_p = volume of a particle

Wetted surface area = NS_p, S_p is the surface area of a particle.

$$R = \frac{eNV_p}{(1-e)NS_p} = \left(\frac{e}{1-e}\right)\frac{V_p}{S_p} \quad \dots \dots \dots \dots (13.6)$$

For a sphere

$$\frac{V_p}{S_p} = \frac{d}{6} \quad \dots \dots \dots \dots \dots \dots \dots \dots (13.7)$$

For non-spherical particles

$$\frac{V_p}{S_p} = K_s \frac{d}{6} \quad \dots \dots \dots \dots \dots \dots \dots (13.8)$$

in which K_s is the shape factor. Also

$$v = \frac{v_s}{e} \quad \dots \dots \dots \dots \dots \dots \dots \dots \dots (13.9)$$

in which v_s is the face or approach velocity. From Equations 13.4, 13.6, and 13.8, we get

$$h_f = \frac{f \ell v_s^2}{8e^2 \left(\frac{e}{1-e}\right) K_s \frac{dg}{6}}$$

$$= f'\left(\frac{\ell}{K_s d}\right)\left(\frac{1-e}{e^3}\right)\frac{v_s^2}{g} \quad \dots \dots \dots \dots (13.10)$$

in which $f' = 6f/8$; f' can also be calculated from the following relationship:

$$f' = 150\left(\frac{1-e}{R_e}\right) + 1.75 \quad \dots \dots \dots \dots \dots (13.11)$$

Equation 13.10, known as the Carman-Kozeny equation, has been derived on the assumption that the bed consists of particles of uniform size. When the bed consists of mixed-size particles:

$$\frac{NV_p}{NS_p} = K_s \frac{d}{6} = \frac{V_{s_1}}{A}$$

i.e. $\quad K_s d = 6 \dfrac{V_{s_1}}{A} \quad \dots \dots \dots \dots \dots \dots (13.12)$

in which A is the total surface of the particles. From Equations 13.10 and 13.12, it can be shown that

$$h_f = f'\left(\frac{\ell}{6}\right)\left(\frac{A}{V_{s_1}}\right)\left(\frac{1-e}{e^3}\right)\left(\frac{v_s^2}{g}\right) \quad \dots \dots \dots (13.13)$$

For particles of uniform size packed homogeneously, the area-volume ratio is

$$\frac{A}{V_{s_1}} = \frac{6}{K_s d}$$

and for the entire bed the average area-volume ratio is given by,

$$\left(\frac{A}{V_{s_1}}\right)_{avg.} = \frac{6}{K_s} \Sigma \left(\frac{p}{d}\right) \quad \ldots \ldots \ldots \quad (13.14)$$

in which p is the fraction of particles by weight retained between adjacent sieve sizes and d is the geometric mean size of the adjacent sieve openings. If the value of $(A/V_{s_1})_{avg.}$ is substituted in Equation 13.1, the head loss in a homogenously packed bed of particles having uniform shape can be found. This equation is applicable to slow sand filters.

In a rapid sand filter the bed is stratified, and therefore some modifications must be made before Equation 13.13 is applied. In a stratified bed the value of the friction factor, f', will not be the same for the entire bed, rather it must be determined for each layer having a uniform size of particles. Assuming that the porosity of the stratified bed is uniform and the particles in each layer are of uniform size,

$$\frac{d}{d\ell} (h_f) = Cf'/d \quad \ldots \ldots \ldots \ldots \ldots \quad (13.15)$$

in which the value of C from Equation 13.10 $= \frac{\ell}{K_s} \left(\frac{1-e}{e^3}\right) \frac{v_s^2}{g}$

$$dh_f = \frac{f'}{d} C\, d\ell$$

If dp is the proportion of particles of size d, then $d\ell = \ell dp$ substituting the value of $d\ell$ in Equation 13.15, we get

$$dh_f = C\ell \quad f' \frac{dp}{d}$$

therefore

$$h_f = C\ell \Sigma \frac{f'p}{d} \quad \ldots \ldots \ldots \ldots \ldots \quad (13.16)$$

Consequently,

$$h_f = \frac{\ell}{K_s} \cdot \frac{1-e}{e^3} \cdot \frac{v_s^2}{g} \cdot \ell \Sigma \frac{f'p}{d} \quad \ldots \ldots \ldots \quad (13.17)$$

Equation 13.17 can be used for a rapid sand filter bed which has stratified layers.

13.7.2 Rose Equations

On the basis of data obtained experimentally, Rose developed an expression to find the head loss through filter beds having uniform spherical or nearly spherical particles. This is similar to the Kozeny equation (Equation 13.10) and can be used directly for hydraulic calculations in rapid sand filters. The equation has been derived as follows:

$$\text{Drag force} \qquad F_d = v_s^2 d^2 D \phi \left(\frac{v_s dD}{\mu} \right) \quad \ldots \ldots \quad (13.18)$$

in which ϕ is read "a function of." When applied to the conditions of flow in filtration, the drag force also equals the difference in pressure (loss of head) between two horizontal planes one grain diameter apart and one grain diameter square in area.

$$\text{Hence} \qquad hDgd^2 \left(\frac{d}{\ell} \right) \propto v_s^2 d^2 D \phi \left(\frac{v_s dD}{\mu} \right) \quad \ldots \ldots \ldots \quad (13.19)$$

$$\frac{h}{\ell} = \phi \left\{ \left(\frac{v_s dD}{\mu} \right) \cdot \left(\frac{v_s^2}{gd} \right) \right\} \quad \ldots \ldots \ldots \ldots \quad (13.20)$$

If a function of porosity is introduced in the equation, we get,

$$\frac{h}{\ell} = \phi \left\{ \left(\frac{v_s dD}{\mu} \right) \cdot \left(\frac{v_s^2}{gd} \right) (e) \right\} \quad \ldots \ldots \ldots \ldots \quad (13.21)$$

in which h_f is the head loss in depth ℓ, v_s is the approach velocity or rate of filtration, d is the characteristic diameter of sand grains, defined as

$$6 V_{s_1} / K_s A$$

Experimental investigation of beds composed of closely graded, smooth spherical particles indicated that

$$\frac{h_f}{\ell} \propto \frac{v_s^2}{gd} \quad \text{and} \quad \frac{h_f}{\ell} \propto \left(\frac{1}{e} \right)^4$$

The above relations can be balanced by a resistance coefficient which is a function of Reynolds Number and is equal to $1.067 \, C_d$. Therefore, the resistance to flow offered by beds of granular material is

$$\frac{h_f}{\ell} = \frac{1.067}{K_s \, e^4} C_d \frac{v_s^2}{gd} = 1.067 \frac{C_d}{g} \left(\frac{v_s^2}{e^4} \right) \frac{A}{6V_{s_1}}$$

$$= 0.178 \frac{C_d}{g} \frac{v_s^2}{e^4} \left(\frac{A}{V_{s_1}} \right) \quad \ldots \ldots \ldots \ldots \quad (13.22)$$

$$\text{Where} \qquad C_d = \frac{24}{R_e} + \frac{3}{\sqrt{R_e}} + 0.34 \quad \text{and} \quad \frac{A}{V_{s_1}} = \frac{6}{K_s d}$$

When the flow is laminar $C_d = \dfrac{24}{R_e}$

Therefore $\dfrac{h_f}{\ell} = \dfrac{1.067}{K_s}\left(\dfrac{24}{R_e}\right)\left(\dfrac{1}{gd}\right)\dfrac{v_s^2}{e^4}$, and $R_e = \dfrac{v_s\,dD}{\mu}$; therefore,

$$\dfrac{h_f}{\ell} = \dfrac{25.6}{K_s}\dfrac{\mu}{g}\dfrac{v_s}{d^2 D e^4}$$

$$= \dfrac{0.711}{K_s}\dfrac{\mu}{gD}\cdot\dfrac{v_s}{e^4}\left(\dfrac{A}{V_{s_1}}\right)^2 \quad \ldots \ldots \ldots \ldots \quad (13.23)$$

The evaluation of the terms in the above equations is straight forward with the exception being d = $(6V_{s_1})/A$. If grains vary in size, they may either be packed homogeneously within the bed or they may be arranged or stratified in order of magnitude from the coarsest to the finest. There is homogeneous packing in slow sand filters, but in rapid sand filters stratification takes place due to backwashing. For homogeneous packing of uniform shaped particles, we have

$$\dfrac{A}{V_{s_1}} = \dfrac{6}{K_s}\int_0^1 \dfrac{dp}{d}$$

in which p = weight fraction of particles of size d.

Assuming that the particles lying between adjacent sieves are substantially uniform we get,

$$\dfrac{A}{V_{s_1}} = \dfrac{6}{K_s}\Sigma\dfrac{p}{d} \quad \ldots \ldots \ldots \ldots \ldots \ldots \quad (13.24)$$

When sieve analyses are used, the term p represents the fraction by weight of particles retained between adjacent sieves. The corresponding diameter of the particles, d, is taken as the geometric mean size = $(d_1 \times d_2)^{\frac{1}{2}}$ of the two sieve openings. Values of shape factor, K_s, for different materials are listed in Table 13.3.

In a rapid sand filter, due to backwashing, stratification takes place. Although the bed consists of nonuniform size particles, it is stratified hydraulically into graded layers consisting of particles of approximately uniform size. Here the largest particles are placed at the bottom of the bed and smallest at the top. If the porosity of the bed is uniform throughout, the loss of head through a layer of thickness $d\ell$ is expressed as:

$$dh = \dfrac{1.067}{K_s}\dfrac{C_d}{g}\dfrac{v_s^2}{e^4 d}\cdot d\ell \quad \ldots \ldots \ldots \ldots \quad (13.25)$$

Therefore, the total head loss is

$$h = \int dh = \dfrac{1.067\,v_s^2}{K_s\,ge^4}\int C_d\dfrac{d\ell}{d}\cdot \quad \ldots \ldots \ldots \quad (13.26)$$

Table 13.3. Shape Factors for Nonspherical Particles[6]

Material	Nature of Grain	K_s
Crushed glass	Jagged	0.65
Pulverized Coal		0.73
Natural Coal Dust (up to $\frac{3''}{8}$ = 9.5 mm)		0.65
Mica Flakes		0.28
Sand, Average for Various Types		0.75
Flint Sand	Jagged	0.66
Flint Sand	Jagged Flakes	0.43
Ottawa Sand	Nearly Spherical	0.95
Wileox Sand	Jagged	0.63
Sand	Rounded	0.82
Sand	Angular	0.73

in which C_d = drag coefficient. Substituting $d\ell = \ell dp$ in Equation 13.23, we get

$$h = \frac{1.067\, v_s^2\, \ell}{K_s g e^4} \Sigma\, C_d\, \frac{p}{d} \quad \cdots \cdots \cdots \cdots \cdots \quad (13.27)$$

Example 13.1

The sieve analysis for a sand to be used in a filter is given in Table 13.4. Calculate the head loss (a) in a 46 cm deep rapid sand filter having a filtration rate of 0.2 cm/sec, (b) in a 46 cm deep slow sand filter having a filtration rate of 0.02 cm/sec. The temperature of the water is 20° C and the porosity of the bed is 0.4. Assume a shape factor for the sand = 0.95.

At 20°C μ/D = 1.004 x 10^{-2} cm²/sec

$$R_e = \frac{vdD}{\mu} = \frac{0.2d}{1.004 \times 10^{-2}} = 19.92\ d$$

$$C_d = \frac{24}{R_e}$$

Using Equation 13.27 for a stratified bed (rapid sand filter)

$$h_f = \frac{1.067\, v_s^2\, \ell}{K_s g e^4}\ \Sigma\, C_d\ \frac{p}{d}$$

$$= \frac{1.067 \times (0.2)^2 \times 46}{0.95 \times 981 \times (0.4)^4} \times 2139.52 = 176\ cm$$

Table 13.4. Results of Sieve Analysis and Summary of Calculations Used in Example 13.1

U.S. Sieve No.	Geometric Mean Diameter, cm $d=\sqrt{d_1 d_2}$	Weight Percentage of Sand Particles Retained p	$\dfrac{p}{d}$	R_e	C_d	$C_d\left(\dfrac{p}{d}\right)$
1	2	3	4	5	6	7
14-20	0.1090	0.12	0.011	2.171	11.05	0.12
20-25	0.0772	0.18	0.023	1.538	15.60	0.36
25-30	0.0646	0.24	0.037	1.287	18.65	0.69
30-35	0.0543	0.31	0.057	1.082	22.18	1.26
35-40	0.0460	3.34	0.726	0.916	26.20	19.02
40-50	0.0354	21.10	5.960	0.705	34.04	202.88
50-60	0.0274	26.40	9.640	0.546	43.96	423.77
60-70	0.0229	23.61	10.310	0.456	52.63	542.62
70-100	0.0177	24.70	13.955	0.353	67.99	948.80
		100.00	40.719			2,139.52

For a mixed bed (homogeneous packing) slow sand filter.

$$h_f = \frac{1.067\, C_D\, v_s^2\, \ell}{g e^4} \cdot \frac{6}{K_s} \Sigma \frac{p}{d}$$

$$\left(\frac{A}{V_{s_1}}\right)_{avg.} = \frac{6}{K_s} \Sigma \frac{p}{d} = \frac{6}{0.95} \times 40.72 = 257,$$

but $d = \dfrac{6}{K_s} \left(\dfrac{V_{s_1}}{A}\right)_{avg.} = \dfrac{6}{0.95} \times \dfrac{1}{257}$

Therefore, $R_e = \dfrac{VdD}{\mu} = \dfrac{(2 \times 10^{-2})}{1.804 \times 10^{-2}} \times \dfrac{6}{0.95 \times 257} = 4.9 \times 10^{-2}$

Therefore, $C_D = \dfrac{24}{R_e} = \dfrac{24}{4.9 \times 10^{-2}} = 490$

$$h_f = \frac{1.067 \times 490 \times (2 \times 10^{-2})^2 \times 46 \times 257}{981 \times (0.4)^4} = 98 \text{ cm} = 0.98 \text{ m}$$

Example 13.2

Calculate the head loss in a rapid sand filter 60 cm deep. The filtration rate is 0.4 cm/sec and the porosity of the bed is 0.4. Sieve analysis for the sand used in the bed is given below. Assume the kinematic viscosity $\mu/D = 1.568 \times 10^{-2}$ cm²/sec and the shape factor $K_s = 0.95$.

ISI Sieve No.	Geometric Mean Size, cm x 100 $100d = 100\sqrt{d_1 \times d_2}$	Percentage of Sand Retained (100p)	$\dfrac{p}{d}$	R_e	C_d	$C_d \dfrac{p}{d}$
140-100	11.2	2.25	0.2	2.86	8.39	1.68
100-60	7.7	10.00	1.3	1.96	12.24	15.91
60-40	5.0	30.50	6.1	1.28	18.75	114.38
40-30	3.5	50.25	14.4	0.89	26.97	388.37
30-15	2.1	7.00	3.34	0.54	44.4	148.30
		100.00	25.34			668.64

Using Equation 13.27 the loss of head

$$h_f = \frac{1.067 \, v_s^2 \, \ell}{K_s \, ge^4} \, \Sigma C_d \, \frac{p}{d}$$

With $e = 0.4, \ell = 60$ cm

$$h_f = \frac{1.067 \times (0.4)^2 \times 60}{0.95 \times 981 \, (0.4)^4} \, (668.64) = 287 \text{ cm} = 2.87 \text{ m}$$

The high rate of filtration and the use of fine sand have caused the high head loss. If v_s is reduced to 0.2 cm/sec, $h_f = 72$ cm.

13.8 HYDRAULICS OF BACKWASHING

Suspended matter is removed as the water flows through the filter bed and it accumulates inside the pores of the bed. As a result, the filter becomes clogged and the head loss increases. To return the filter to an efficient mode of operation, it must be cleaned by backwashing, which scours out the matter collected inside the pore spaces. When water flows through the bed from the bottom towards the top, filter grains are lifted, and expansion of the sand bed takes place. In order to wash the filter properly, a velocity should be reached at which the particles no longer remain stationary and in contact with one another but are fluidized with all the grains in suspension. When the bed becomes fluidized, the maximum resistance offered by the particles to the passing fluid is their effective weight (weight of the particles in water). Further expansion of the bed and higher wash rate do not increase the friction loss and the cleaning efficiency is unchanged. The head loss through the expanded bed can be calculated as follows:

Pressure of water tending to lift the particles $= Dgh_e$ . . . . . (13.28)

Effective weight of
grains per unit area $= (D_p - D) g \left[\dfrac{\text{Volume of grains}}{\text{Area}} \right]$

$$= (D_p - D) g \left[\dfrac{\text{Total Volume } (1 - e_e)}{\text{Area}} \right]$$

$$= (D_p - D) g \left[\dfrac{\text{Area} \, \ell_e \, (1 - e_e)}{\text{Area}} \right]$$

$$= (D_p - D) g. \, \ell_e \, (1 - e_e) \quad \dots \dots \dots (13.29)$$

Therefore, $Dgh_e = (D_p - D) g. \ell_e (1 - e_e)$

$$h_e = \frac{(D_p - D)}{D} \ell_e \, (1 - e_e) \quad \dots \dots \dots \dots (13.30)$$

in which

D = density of the fluid, kg/m^3
D_p = density of the particles, kg/m^3
h_e = head loss in the fluidized bed, m
ℓ_e = depth of bed after expansion, m
e_e = porosity of the expanded bed
g = acceleration due to gravity m/sec^2

According to Fair and Geyer,[7]

$$\left(\frac{v_s}{v} \right)^2 = \phi \, e_e$$

Where v_s and v are terminal settling velocity of the particles and the face or approach velocity of wash water respectively.

For the hydraulic expansion of a bed of uniform sand grains, the following relationship has been established experimentally.

$$\left(\frac{v_s}{v} \right)^2 = \left(\frac{1}{e_e} \right)^9 , \text{ or}$$

$$e_e = \left(\frac{v}{v_s} \right)^{0.22} \quad \dots \dots \dots \dots \dots \dots (13.31)$$

As the total volume of particles per unit area remains constant, the following expression for a bed of uniform grain size can be deduced.

$$\ell (1 - e) = \ell_e \, (1 - e_e)$$

$$\text{or } \ell_e = \frac{\ell(1-e)}{(1-e_e)} = \frac{\ell(1-e)}{\left[1 - \left(\dfrac{v}{v_s}\right)^{0.22}\right]} \quad \ldots \ldots \quad (13.32)$$

in which

ℓ = depth of unexpanded bed, m

e = porosity of the unexpanded bed

In a stratified bed of mixed sized particles, the bed will be fluidized when

$$v > v_s \, (e_e)^{4.5} \quad \text{and for such beds,}$$

$$\frac{\ell_e}{\ell} = (1-e) \, \Sigma \frac{p}{(1-e_e)} \quad \ldots \ldots \ldots \quad (13.33)$$

where p = weight fraction of the particles (percentage of sand) in a layer. In the above equation it is assumed that the particles lying between adjacent sieves are substantially uniform.

Example 13.3

A bed of 60 cm depth is washed at a rate of 1.1 cm/sec. The kinematic viscosity of water is $1.568 \times 10^{-2}\, cm^2/sec$. The porosity of the stratified bed is 0.4.

The grain size distribution is given below:

ISI Sieve No.	% of Sand Retained
140 - 100	2.25
100 - 60	10.00
60 - 40	30.50
40 - 30	30.25
30 - 15	7.00
Total	100.00

Calculate: 1. Depth of the expanded bed, and
2. The head loss through the bed.

ISI Sieve Numbers	Weight Fraction of Particles Retained p	Geometric Mean Diameter, cm d	Terminal Settling Velocity v_s cm/sec	Porosity of Expanded Bed e_e	$\dfrac{p}{1-e_e}$
1	2	3	4	5	6
140-100	0.0225	0.112	14.70	0.565	0.052
100-60	0.1000	0.077	9.64	0.610	0.266
60-40	0.3050	0.050	5.86	0.699	0.997
40-30	0.5025	0.035	3.90	0.766	2.140
30-15	0.0700	0.021	2.18	0.880	0.506
					3.961

The settling velocity shown in Column 4 is computed from Stokes Law.

$$v_s = \sqrt{\frac{4}{3} \frac{g}{C_d} \left(\frac{D_p - D}{D}\right) d} = \sqrt{\frac{4}{3} \times \frac{981}{C_d} \cdot \left(\frac{2.65 - 1}{1}\right) d}$$

$$= 46.46 \sqrt{\frac{d}{C_d}}$$

From Equation 11.3, assuming spherical particles

$$C_d = \frac{18.5}{(R_e)^{0.6}} = \frac{18.5}{\left(\frac{dv_s D}{\mu}\right)^{0.6}} = \frac{1.529}{d^{0.6} v_s^{0.6}}$$

Substituting this value of C_d in the above,

$$v_s = 177.76 \, d^{1.143}$$

Hence for the largest particle, $v_s = 177.76 \, (0.112)^{1.143} = 14.7$ cm/sec.
Column 5: Substituting $v = 1.1$ cm/sec and v_s for the largest particle in Equation 13.31, we get

$$e_e = \left(\frac{1.1}{14.7}\right)^{0.22} = 0.565$$

Column 6: Substituting the value of e_e calculated above,

$$\frac{p}{1 - e_e} = \frac{0.0225}{1 - 0.565} = 0.052$$

From Equation 13.30, head loss

$$h_e = \ell_e \, (1 - e_e) \left(\frac{D_p - D}{D}\right)$$

$$= 60 \, (1 - 0.565) \left(\frac{2.65 - 1}{1}\right) = 43 \text{ cm}$$

Expanded depth is calculated from Equation 13.33

$$\ell_e = \ell (1 - e) \, \Sigma \, \frac{p}{1 - e_e}$$

$$\ell_e = 60 \, (1 - 0.4) \times 3.961 = 143 \text{ cm}$$

13.9 UNDERDRAINS OF FILTERS (Figure 13.4)

A.   Perforated pipe underdrain system
    a.   Manifold: The main manifold (header) can be circular or rectangular in cross section depending upon the materials of which it is constructed.

Materials Used for Manifold

1.   Steel. It has the following advantages:
    i.   Easy to install.
    ii.   Easy to carry or store.
    iii.   Not subjected to cracking or breakage.
    iv.   Manufactured in small sections which are easily welded together.
   It has the following disadvantages:
    i.   Subject to corrosion.
    ii.   Expensive and in some countries not locally manufactured.
2.   Asbestos Cement. It has the following advantages:
    i.   Generally locally produced.
    ii.   Not subject to corrosion.
   It has the following disadvantages:
    i.   Subject to cracking and breakage.
    ii.   Heavy and difficult to handle.
3.   Concrete. It has the following advantages:
    i.   May be cast *in situ.*
    ii.   Inexpensive and does not need to be transported.
   It has the following disadvantages:
    i.   Skill required to form and place.
    ii.   Heavy and difficult to handle.
4.   Cast Iron. It has the following advantages:
    i.   Less subject to corrosion than steel.
    ii.   Resists breakage better than asbestos cement.
   It has the following disadvantages:
    i.   Subject to cracking and breakage but less than asbestos cement.
    ii.   Heavy and difficult to handle.
    iii.   Expensive.
    b.   Laterals

Materials Used for Laterals

1.   Copper. Advantages:
    i.   Light and easy to handle.
    ii.   Resists corrosion.
   Disadvantages:
    i.   Very expensive.
2.   Asbestos Cement. Advantages:
    i.   Easy to drill and cut.
    ii.   Usually locally available.

       iii.   Inexpensive.

       iv.   Resists corrosion.

    Disadvantages:

       i.   Subject to cracking and breakage.

  3.   Steel. Advantages:

       i.   Easy to handle and cut.

       ii.   Easily welded to the manifold.

    Disadvantages:

       i.   Subject to corrosion.

       ii.   Relatively expensive.

B.   Tile underdrain system

  1.   Concrete Tile. Advantages:

       i.   Inexpensive and can be locally produced.

    Disadvantages:

       i.   Sometimes joints fail.

       ii.   Installation requires skilled workers.

  2.   Vitrified Clay Tile. Advantages:

       i.   Inexpensive.

       ii.   Easy to assemble.

    Disadvantages:

       i.   Subject to cracking.

  3.   Plastic Nozzles. Advantages:

       i.   Inexpensive.

       ii.   Gives excellent results.

    Disadvantages:

       i.   Installation requires skilled workers.

13.10 DESIGN CRITERIA FOR RAPID SAND FILTERS

A.   Filter bed

  1.   The area of the bed is between 20 and 30 m^2.

  2.   The dimensions of the bed may be between 2.7 and 6 m.

  3.   The ratio of length to breadth is between 1.25 and 1.33.

  4.   At least two filter beds must be provided.

  5.   Total depth (gravel and sand) varies from 1 to 1.2 m.

  6.   Free board is 30 to 45 cm.

  7.   Depth of water above the sand bed is 1 to 1.5 m.

  8.   Height of underdrainage system is 45 to 60 cm.

  9.   Rate of filtration varies between 4 and 5 $m^3/m^2/hr$.

B.   Backwashing

  1.   Backwashing time is between 3 and 10 minutes (usually 5 min).

  2.   Rate of backwashing may be between 25 and 37 $m^3/m^2/hr$.

  3.   Compressed air for cleaning may be between 1 and 1.5 m^3 of free air/m^2/min.

  4.   Pressure of backwashing is between 1 and 2 atmospheres.

C.   Underdrains
   1.   Backwashing pressure at the laterals or strainers should not be more than 1 atmosphere.
   2.   The velocity of flow inside laterals should not be more than 2 m/s.
   3.   The spacing of laterals should be 15 to 30 cm.
   4.   Maximum length of the lateral is 60 times its diameter.
   5.   The size of the openings in the laterals is 6 to 12 mm.
   6.   The spacing of openings is 7.5 cm for 6 mm diameter openings and 20 cm for 12 mm diameter openings.
   7.   The total area of the opening should be about 0.3 to 0.5 percent of the surface area of the filter bed.
   8.   The cross sectional area of the laterals ranges between twice the area of the openings of 12 mm diameter and four times the area of openings of 6 mm diameter.

Example 13.4

Design a rapid sand filter for a flow of 400 m^3/hr (app. 2000 gpm). Assume the rate of flow through the filter bed = 5 $m^3/hr/m^2$.

Area of the bed $= \dfrac{400}{5} = 80 \ m^2$

Use four filter beds each $\dfrac{80}{4} = 20 \ m^2$

Dimensions of each filter = 4 x 5 m

Laterals and Header (Figure 13.3d)

Size of the openings in a lateral = 6 mm to 12 mm
Cross sectional area of the lateral   =   four times the area of the openings = 4 A_p (if the size of the opening is 6 mm)

$$A = \frac{\pi d^2}{4}$$

$\therefore$ $A = 4 \times \dfrac{\pi \ 36}{4} = 113.1$

But $A = 2 \ A_p$ (if the size of opening is 12 mm)

$\qquad = 2 \times \dfrac{\pi \ 144}{4} = 226.2$

Use $A = 226.2 = \dfrac{\pi}{4} \ d^2$

$\qquad \therefore$ d (diameter of the lateral) = 17 mm

Number of laterals $= \dfrac{500}{20} = 25$ on each side of manifold

Total number of laterals = 50

Total cross sectional area of laterals $= \frac{226.2}{100}$ x 50 = 113 cm^2

Total area of the header = (1.75 to 2) x A = 2 x 113 = 226 cm^2

Diameter of the header = $\sqrt{\frac{226 \times 4}{\pi}}$ = 17 cm

Total length of the lateral = 400 - (2 x 17 + 3.5 + 5 + 3.5) = 354 cm

Using two headers, length of each lateral = $\frac{354}{4}$ = 88 cm

The length of the lateral should be less than 60 times its diameter, 60 x 1.7 = 102 cm > 88 cm. Hence the condition is satisfied.

<u>Check</u>: Area of openings in laterals should be 0.2 to 0.3 percent of filter area.

Area of opening provided $= \frac{\pi}{4}$ x 12^2 = 113 mm^2 ; assume spacing of openings

= 17 cm

Therefore, number of openings in each lateral = 5

Total area of openings in laterals $= \frac{2 \times 50 \times 5 \times 113}{100}$ = 565 cm^2

$\therefore$ Percent area of openings $= \frac{565}{20 \times 100^2}$ x 100 = 0.282 < 0.3%, hence
satisfactory

<u>Backwashing</u>

Assume rate of backwashing = 35 m^2/hr/m^2.
Assume that two filter beds are washed at the same time.
Wash area = 20 x 2 = 40 m^2.
Quality of water needed for backwashing = 40 x 35 = 1400 m^3/hr.
Head of wash water = 10 m.
Frictional resistance = 4 m (app.).
$\therefore$ Total head required = 14 m.
Use two pumps in parallel each having a capacity of 700 m^3/hr against a head
of 14 m.
Instead of pumps elevated tanks may be used.
Use one elevated tank to supply backwash water for 5 minutes.

Volume of the elevated tank $= \frac{1400}{60}$ x 5 = 120 m^3.

Size of the tank = 4 x 5.5 x 5.5 m.
Height = 14 m.
If pumps are to be used during backwashing,

Volume of the storage tank $= \frac{120}{2}$ = 60 m^3

Size of the tank = 3 x 4.5 x 4.5 m

To fill the elevated tank with treated water, either use a branch from the pumping main or use a special pump (duration of pumping 1 to 4 hours).

Trough

$$\text{Flow through the trough} \quad Q = 2.49 \, bh^{3/2} \, (\text{m}^3/\text{s})$$

$$\frac{400}{3600} = 2.49 \, b \, (0.3)^{1.5}$$

$$b = 0.57 \, \text{m} = 57 \, \text{cm}$$

$$\text{Total depth} = 0.3 + 0.05 \, (\text{free board})$$

$$= 0.35 \, \text{m}$$

$$\textbf{Trough dimension} = 57 \times 35 \, \text{cm}$$

14. Disinfection

14.1 GENERAL

Water is disinfected to destroy the pathogenic organisms. Microorganisms are removed from water in various degrees during the process of settling, addition of chemicals and filtration, but to make the water safe for human consumption it must be disinfected. Chlorine gas and chlorine compounds (e.g. chloride of lime or calcium hypochlorite), which are relatively cheap and have prolonged action as disinfectants, are commonly used. Apart from the germicidal effect of chlorine, it also destroys taste and odor producing compounds, algae, and related water blooms and slime organisms. It also helps in the oxidation of iron, manganese, and H_2S.

14.2 THEORY OF DISINFECTION

Disinfection of water kills bacteria, protozoa, and viruses, and the quantity of disinfectant required is small and not toxic to humans. If the other factors remain constant, the killing of organisms is proportional to the time of contact and concentration of the disinfecting agent. In practice, however, the rate of kill may actually increase or decrease with time. The presence of interferences, variations in the resistivity of cells, among other factors, may contribute to a decrease in the rate. Increases may be attributed to the dependence of kill rate on the kind and concentration of disinfectant within the cell. The characteristics of the water being treated has a significant influence on disinfection processes. Chlorine is the most commonly used disinfectant for water, and the chemical reactions of chlorine are discussed in Section 14.5.

14.3 ACTION AND PURPOSE OF DISINFECTION

Any disinfectant should kill all pathogenic organisms. The reaction should be complete under normal conditions, e.g. temperature, flow, quality of water, and time available. It should make the water neither toxic nor unpalatable, and should be easy to handle and economical as well. It should remain as residue for a safe period so that any likelihood of further contamination is eliminated.

219

14.4 DISINFECTING AGENTS

14.4.1 Physical Agents

Boiling water kills disease causing organisms in 15 to 20 minutes, although for safety water should be boiled for a longer period. Sunlight is a natural disinfectant because of ultra violet rays.

14.4.2 Chemical Agents

Chlorine, bromine, and iodine, which belong to the Halogen group of chemicals, are effective disinfectants. Oxidizing agents such as potassium permanganate, chlorine dioxide, and ozone may also be used as disinfectants. Ultraviolet light is an effective disinfectant.

14.5 DISINFECTION BY CHLORINE

Chlorine is employed in the form of hypochlorites or as free chlorine. But free residual chlorine is not molecular chlorine free to act as dissolved gases except at a pH value of 5 or below. Chlorine combines with water at pH 5 and 6, to form hypochlorous and hydrochloric acids.

$$Cl_2 + H_2O \rightleftharpoons HOCl + HCl \quad \ldots \ldots \ldots \ldots (14.1)$$

Hydrochloric acid is neutralized by the alkalinity, and hypochlorus acid becomes dissociated.

$$HOCl \rightleftharpoons H + OCl^- \quad \ldots \ldots \ldots \ldots (14.2)$$

The quantities of HOCl and OCl^- ion depend upon the pH value. At a pH value of 8.5, 90 percent of HOCl is ionized to hypochlorite ion. Cl_2 in the elemental form remains for a short while in water. HOCl and OCl remain in water and are called free available chlorine.

HOCl (hypochlorous acid) is a more effective disinfectant. Most waters have a pH value between 6 and 7.5, and therefore 40 to 95 percent of the free residual Cl_2 is in the form of HOCl. Softened water with a high pH value must be treated with higher doses of chlorine to compensate for the low disinfecting power of OCl^-.

The reaction with calcium hypochlorite is as follows:

$$Ca(OCl)_2 + H_2O \rightleftharpoons Ca^{++} + H_2O + 2OCl^- \quad \ldots (14.3)$$

$$H^+ + OCl^- \rightleftharpoons HOCl \quad \ldots \ldots \ldots \ldots (14.4)$$

Ammonia or organic nitrogen, if present in water, reacts with chlorine and hypochlorous acid to form monochloramine (NH_2Cl), dichloramine ($NHCl_3$) and nitrogentrichloride (trichloramine, NCl_3),

$$NH_3 + HOCl \longrightarrow NH_2Cl + H_2O \text{ monochloramine } \; . \; . \; . \; (14.5)$$

$$NH_2Cl + HOCl \longrightarrow NHCl_2 + H_2O \text{ dichloramine } \; . \; . \; . \; . \; (14.6)$$

$$NHCl_2 + HOCl \longrightarrow NCl_3 + H_2O \text{ trichloramine } \; . \; . \; . \; . \; (14.7)$$

The above reactions also depend upon the pH value of water. The monochloramine and dichloramine are active disinfectants and are called "combined available chlorine."

Chlorine reacts with various substances and especially with reducing agents.

$$H_2S + 4Cl_2 + 4H_2O \longrightarrow H_2SO_4 + 8HCl \; . \; . \; . \; . \; . \; (14.8)$$

Organic compounds also react with chlorine and thereby increase its consumption. Chlorine will be effective as a disinfectant only after the side demand reactions mentioned above are satisfied.

14.6 METHODS OF CHLORINATION

14.6.1 Prechlorination

Chlorine may be added directly to raw water. The bacteria are killed during prechlorination minimizing the possibility of bacteria passing through the filter bed. Prechlorination improves coagulation and reduces taste and odor by oxidizing organic matter.

The growth of algae and other organisms which quickly choke a filter must be controlled resulting in an increase in the length of filter run. The dose should be such that prior to filtration a residual of 0.1 to 0.2 mg/ℓ should be present (generally 5 to 15 mg/ℓ). However, post chlorination should not be abandoned regardless of the advantages of prechlorination.

14.6.2 Post Chlorination

In post chlorination, chlorine is added to the filtered water (Figure 13.5). Chlorine may be added to the outlet channel or the suction line of the pump. The chlorine dose (2 to 3 mg/ℓ in general) depends upon the nature of water and the contact period required.

14.6.3 Break Point Chlorination

As has been discussed earlier, the chlorine demand of water depends upon the organic and inorganic substances present. As the dose of chlorine is increased, combined available residual also increases. This will continue until the residual starts decreasing, indicating the oxidation of chloramines and other chloroorganic compounds (Figure 14.1). When the oxidation is complete, there will be a further rise in the chlorine residual, and chlorine will

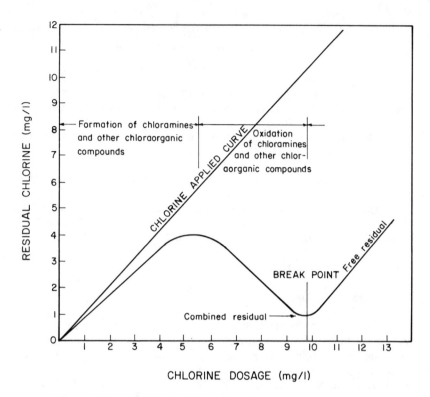

Figure 14.1. Residual chlorine curve showing break point.

be in the form of free available chlorine. The point at which this takes place is called the "break point." At and beyond break point all odor and taste disappear and a high germicidal effect is achieved. Dosages of chlorine generally required in the treatment of drinking water are given in Table 14.1.[1]

14.7 CHEMICALS USED IN DISINFECTION

Water is usually disinfected with chlorine gas or liquid chlorine or chlorine compounds such as bleaching powder $(CaOCl_2)$ and calcium hypochlorite $Ca(OCl)_2$. Chlorine available in bleaching powder varies between 33.5 and 39 percent and in calcium hypochlorite it is about 70 percent.

Liquid chlorine is chlorine gas which is compressed to the point at which it liquifies. It is usually available in cylinders, which are fitted to an assembly called a chlorinator (Figure 14.2). With the help of manometers and control valves, the dosage of chlorine can be adjusted.

Table 14.1. Minimum Chlorine Dosages Generally Required For Drinking Water at 20°C[1]

pH Value	Minimum Concentration of Residual Chlorine mg/ℓ (Disinfecting Period at Least 10 min)	Minimum Concentration of Combined Residual Chlorine mg/ℓ (Disinfecting Period at Least 60 mm)
6-7	0.2	1.0
7-8	0.2	1.5
8-9	0.4	1.8
9-10	0.8	Not recommended
10-11	0.8 + higher dose	Not recommended

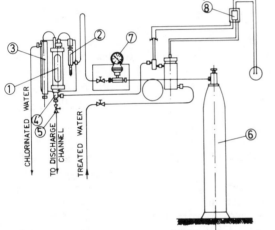

1 _ CHLORINATED WATER METER
2 _ FILTER
3 _ MIXING PIPE
4 _ WATER CONNECTION
5 _ OUTLET VALVE
6 _ CHLORINE CYLINDER
7 _ MANOMETER FOR CHLORINE
8 _ RELAY

Figure 14.2. Chlorinator (space required 2.0 x 1.0 x 2.3 m high).

14.8 FACTORS INFLUENCING CHLORINATION

The following are some of the factors which influence chlorination.
1. Suspended solids in water may shield the bacteria from chlorine.
2. The disinfecting power is reduced due to the presence of organic matter.
3. Chlorination is more effective in waters of low alkalinity and low pH values.
4. The effectiveness of chlorine is reduced due to the presence of nitrites, iron, and manganese which react with chlorine.

14.9 DESIGN OF CHLORINE CONTACT TANKS

Problems associated with the design of chlorine contact tanks stem from the fact that most designs are based on a theoretical detention time determined by dividing the flow rate by the tank volume. In practice, the actual detention times may vary between 30 and 80 percent of the theoretical detention times. Shorter residence times are caused by dead spaces and short-circuiting with decreased chlorination efficiencies, and chlorine concentrations must be increased to produce desired degrees of disinfection. Increasing the concentration of chlorine often has serious drawbacks. As well as being an inefficient way to utilize the disinfectant properties of chlorine, it also increases operational costs. Increasing the chlorine dose causes excessive wear of equipment, because of corrosion resulting from the contact of equipment with high chlorine concentrations.

Short-circuiting has another effect on adequate operations of chlorine contact tanks. With short-circuiting, residence times may be continually changing. This causes difficulty in maintaining prescribed levels of chlorine residuals. The frequent attention of an operator is required to alter chlorine doses in maintaining constant chlorine residuals.

To provide adequate disinfection of wastewater, the basic approach to good contact tank design should include a thorough investigation of hydraulic characteristics of various designs and then the selection of design features which will optimize hydraulic performance. Some important design considerations include optimization of mixing, contact time, and chlorine dose.

14.9.1 Evaluation of Hydraulic Characteristics

The hydraulic characteristics of a chlorine contact tank are generally determined by conducting tracer studies on flow patterns through the tank. Several possible tracers are available. Salt is a common tracer and has been used to determine detention times in contact tanks; however, it is often difficult to handle the large amounts of salt generally required for such studies. Radioactive tracers are another possibility; however, these are almost never used because of the hazard and regulations controlling their release.

Perhaps the most useful tracers are fluorescent dyes. Most of these are inexpensive and easy to obtain. Two of the dyes commonly used in contact tank tracer studies are Rhodamine WT and Rhodamine B. Other fluorescent dyes are also available, and the choice of which dye to use is a matter of personal judgment. The Rhodamine dyes, however, offer the advantages of being detectable at very low concentrations and having low sorption tendencies. Also, turbidity has very little effect on the response of the dye. Fluorescence of the dyes at concentrations as low as 0.01 ppb can be detected with a fluorometer.

In conducting tracer studies the dye or other tracer should be injected into the contact tank at about the same point at which the chlorine solution would enter the tank. If possible, the tracer should also be injected below the water surface to avoid scattering of the tracer by wind on the surface. The most desirable method of conducting tracer studies is to obtain a continuous record of the tracer concentrations at the tank outlet. If fluorescent dyes are used, this may be done by using a continuous flow fluorometer connected to a recorder. This type of approach is more reliable than the collection of grab samples.

The flow characteristics of the contact tank may be determined by evaluating the data obtained from tracer studies in one of several ways. The methods include conventional, statistical, and dynamic analyses.[2] Conventional and statistical analyses are the most commonly used. The dynamic approach is basically a mathematical modeling technique and will not be discussed.

The conventional method of analysis consists of selecting specific points from the dispersion flow curve as indices to describe the performance characteristics of a tank. The points and indices commonly used are described as follows:

T = Q/V (theoretical detention time)

t_i = time for tracer to initially appear at the tank outlet

t_p = time for tracer at outlet to reach peak concentration

t_{10}, t_{50}, t_{90} = time for 10, 50, and 90 percent of the tracer to pass at the outlet of the tank

t_g = time to reach the centroid of the effluent curve

t_i/T = index of short-circuiting

t_p/T = index of modal detention time

t_{50}/T = index of mean detention time

t_g/T = index of average detention time

t_{90}/t_{10} = Morrill Dispersion Index—indication of degree of mixing

In constructing disperion flow curves, it is common practice to use dimensionless expressions for tracer concentrations and times. This is done to facilitate comparisons of hydraulic performance between tanks where different tracer concentrations and detention times are involved. The dimensionless dispersion flow curve is obtained by plotting C/C_0 against t/T where C is the tracer concentration at any time t, C_0 is the initial tracer concentration, and T is the theoretical detention time (Q/V). A typical dispersion flow plot is presented in Figure 14.3.

The parameter which is probably the most useful in accurately describing hydraulic performance is the Morrill Index (MI).[3] As the MI approaches 1.0, the flow through the tank approaches ideal plug flow. The larger the MI, the more closely the flow in the tank approaches backmix (complete mixed) reaction conditions. The two extreme flow conditions are displayed in Figure 14.4.

There are several different statistical approaches used to evaluate hydraulic performance. One approach, which has gained widespread acceptance, describes the flow regime of a basin in terms of plug flow and perfect mixing.[3,4] It also uses descriptive parameters to define effective space and dead space. A variation of this approach uses the entire tracer curve

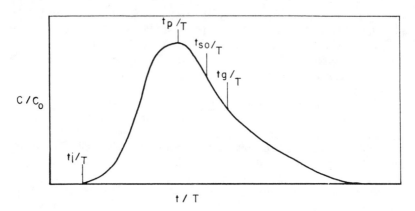

Figure 14.3. Typical dispersion flow curve.

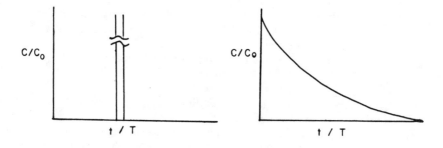

Figure 14.4. Comparison of plug and backmix flow.

to describe hydraulic efficiency in terms of a function of time, $F(t)$.[5] This function is calculated from the following equation

$$\text{Log}\,[1 - F(t)] = [-\text{Log}\,e/(1 - p)\,(1 - m)]\,[t/T - p(1 - m)] \quad . . .(14.9)$$

m	=	dead space fraction
1-m	=	effective fraction
p	=	plug flow fraction
1-p	=	perfect mixing fraction
t	=	any time corresponding to time used to get F(t)
T	=	theoretical detention time

Probably the most widely used statistical approach is the chemical engineering dispersion index. It is considered to be extremely reliable, since it is calculated using the entire dispersion flow curve. The dispersion index, δ, is calculated from the following equation.[3]

$$\delta = \sigma^2 = \frac{\delta t^2}{tg^2} \quad \ldots \ldots \ldots \ldots \ldots \quad (14.10)$$

$$\sigma t^2 = \left(\frac{\Sigma t^2 c}{\Sigma c}\right) - \left(\frac{\Sigma tc}{\Sigma c}\right)^2 \quad \ldots \ldots \ldots \quad (14.11)$$

$$tg = \frac{\Sigma tc}{c} \quad \ldots \ldots \ldots \ldots \ldots \quad (14.12)$$

In these equations, c is the tracer concentration at any time t, $\delta = \sigma^2$ is equal to the variance of the flow-through curve.

The dispersion index has the strongest statistical probability of correctly describing the hydraulic performance because it includes all points on the dispersion flow curve. Conventional parameters use only one point, or at the best, only a portion of the curve. In comparing the dispersion index with conventional parameters, it has been found that the Morrill Index is closely correlated with the dispersion index and can be considered as the most reliable conventional parameter in accurately describing the hydraulic performance of a tank. The least reliable indicators of flow characteristics are considered to be the percent of effective space, t_{50}/T, and t_i/T.[3]

14.9.2 Elements of Contact Tank Design

The primary objective of good chlorine contact tank design is to design for hydraulic performance which will allow for a minimum usage of chlorine with a maximum exposure of microorganisms to the chlorine. An evaluation of a number of chlorine contact tanks indicates that mixing, detention time, and chlorine dosage are the critical factors to consider in providing adequate disinfection.

Initial mixing of the chlorine solution with water is necessary for providing uniform contact of chlorine with microorganisms and for preventing chlorine stratification in the contact tank. This is especially important because most of the disinfection takes place within the first few minutes of contact. It is also important to note that most of the chlorine demand is exerted during this same period. Since the formation of chloramines in water is extremely rapid, it must be remembered that free chlorine is much more effective as a disinfectant than chloramines. Chloramines are ineffective in killing viruses in comparison with free chlorine.

Mixing can be accomplished by applying the chlorine solution to the water either in a pressure conduit under turbulent conditions or with a mechanical mixer. A turbulent reactor is generally considered to be the most effective in producing maximum bacterial kill in the shortest contact time. It has been found that a contact time of 0.1 to 0.3 minutes is generally sufficient in a turbulent reactor. Slightly longer might be required when a mechanical mixer is used. If a mechanical mixer is used, the chlorine solution should be added to the water immediately upstream from the mixer. Another form of mixing, which has been found to be effective, is the use of a hydraulic jump in combination with over and under baffles. Both the turbulent reactor and the baffle system of mixing offer the advantage of reducing operation and maintanence costs over those for the mechanical mixer.

Rapid mixing is followed by flow of the chlorinated water into the contact tank. Most approaches to good contact tank design are based on the idea that plug flow is the most desirable hydraulic performance characteristic to achieve in producing efficient disinfection. Plug flow decreases short-circuiting, dead spaces, spiraling, and eddy currents and also closes the gap between theoretical and actual detention times. However, not all designs are based on the objective of achieving plug flow. The use of a series of backmix reactors to improve chlorination efficiency has been proposed. In this approach, the tank shapes are not important as long as stratification and short-circuiting are eliminated. One advantage to this approach is the ease with which treatment capacity could be increased by simply adding another reactor. However, high initial and operational costs usually offset this advantage.

For the design of tanks in which plug flow is the objective, tank shape is an important consideration. Ideally, plug flow conditions could best be achieved by using a long, narrow, straight contact chamber. A pipe, for example, would be a good contact chamber. However, because of cost and space limitations, this approach is generally not practicable. Circular contact tanks have been used, but generally they do not perform efficiently with respect to hydraulic characteristics. Most tanks are based on a rectangular shape, which is the most practical design.

Conventional design practices can be enhanced by paying particular attention to inflow and outflow structures. They should be designed in such a fashion as to distribute flow uniformly across the tank cross-section. One of the most effective designs is that of a sharp-crested weir covering the width of the contact tank at the inlet and outlet. This design minimizes the weir overflow rate and greatly enhances hydraulic characteristics through the tank.

A common practice for improving plug flow conditions in a contact tank involves the use of baffles. Longitudinal baffles are generally more effective than cross baffles. In a study of seven different types of chlorine contact tank configurations, it was found that the longitudinally baffled serpentine flow and flow resulting in an annular ring around a secondary clarifier were the best configurations for approaching ideal plug flow. Both have the effect of increasing the ratio of the length to the width (L/W) of the contact tank. The L/W ratio is often considered to be the most important design consideration for chlorine contact tanks. A minimum L/W ratio of 40:1 is recommended in order to obtain maximum plug flow performance. Baffles have also been used effectively across the width of contact tanks. Hydraulic performance has been improved by placing baffles near the inlet end of tanks to suppress the kinetic energy of incoming jets.

Often, baffles by themselves are not sufficient to produce desired hydraulic characteristics. Hammerhead shapes at baffle tips have been demonstrated to reduce short-circuiting and flow separation. Corner fillers have been used to eliminate dead spaces and to decrease the buildup of solids in corners. These fillers, however, seem to have little effect on flow characteristics. In some cases, directional vanes around the ends of baffles have been found to produce lower head losses and to produce more uniform flow through the contact tank.

One final design consideration is that of depth. In very shallow contact tanks, it is possible for wind to cause short-circuiting. However, this is generally not a problem in tanks designed with standard design depths of approximately 3 m.

For existing chlorine contact tanks, it is generally not possible to completely redesign the tank. However, improvements can be made in flow characteristics with practical alterations. Gates added to screen and sludge notches have been found to reduce short-circuiting. Spiraling flow patterns have been eliminated by circular baffle plates placed at tank inlets. Additional improvements can be made by using directional vanes to direct flow in a more uniform fashion and by using stop baffles with curved vanes to reduce eddying. Another possible way of improving hydraulic performance in existing tanks is to use pre-cast baffles. These can be installed with minimum down time. Although it is more efficient to use longitudinal baffles, it may be more economical to use cross baffles. It has been demonstrated that baffles installed in a maze configuration improved performance sufficiently to make economical factors more important in choosing a design than efficiency considerations.[6]

In conclusion, the most important design considerations for efficient disinfection appear to be rapid and complete initial mixing, an adequate L/W ratio to produce near plug flow conditions, and a sufficiently long residence time to produce an optimal amount of disinfection for the chlorine dose applied with a minimum amount of chlorine residual remaining in the effluent.

15. Taste and Odor Control

15.1 TYPES OF TASTES AND ODORS

Tastes and odors are each divided into four basic groups. Tastes consist of sweet, sour, bitter, and salty. The odor divisions are sweet, sour, burnt, and goaty. Tastes and odors are so closely connected that there appears to be an unlimited number of tastes while, in truth, only the four basic ones exist. With the olfactory receptors closed, it has been shown that sensations normally associated with tastes are actually odors resulting from chewing and drinking actions.

Sour, salt, and sweet tastes are not generally found in water complaints. This is due to their occurring in low concentrations at which their tastes are not detectable. The bitter taste is detectable in very low concentrations, and therefore is more of a problem in water supplies, as most substances dealt with in water treatment are in very low concentrations. Chlorination may convert small quantities of the aromatic series to a bitter substance, which results in tastes.

Listing only four basic odors might tend to confuse, but if it is realized there are limitless combinations of varying concentrations possible, the many odors recognized by man can be understood. For a substance to produce an odor: (a) it must be volatile, (b) lipoid solubility is essential, (c) and it must be enhanced by insaturation.

15.2 CAUSES OF TASTES AND ODORS

Algae, decaying vegetation, and industrial wastes are the usual causes given for tastes and odors in water supplies.[1] The odor-producing algae were considered to cause most of the tastes and odors in water by releasing volatile oils during the growth phase and by decomposition upon death. Silvey[2] has produced information which attributes many of the causes of tastes and odors associated with algae to the actinomycetes. Industry has been credited with the production of all phenolic tastes and odors, but it has been shown that vegetation and domestic sewage produce considerable phenolic compounds.[3] Also bacteria, protozoa, metazoa, and weeds contribute to odors and tastes in water.

15.3 PREVENTIVE TREATMENT

Control of algae growths is the primary factor in preventive treatment. Microscopic examination to count and identify the organisms is the first step in control work.

Various points throughout the reservoir system should be sampled. Depth of water, exposure to sunlight, conditions of raw water supply, and numerous other small variations will produce surprising differences in type and number of organisms present. The temperature of the water should be watched closely. Hydro-biologists have found that critical temperatures for algae and protozoa lie close to 41°, 50°, 59°, 68°, and 77° F (5°, 10°, 15°, 20°, and 25°C). As the temperature passes the critical point, certain species are stimulated to activity. Temperature changes can be rapid in shallow areas of reservoirs and will produce unexpected organisms.

When to treat a reservoir can only be determined by close observation of the water temperature and number and type of organisms present. The upper limit of each type algae should be determined for a particular reservoir, and when this limit is reached, a treatment should be prescribed.

The quantity of copper sulfate necessary for treatment will vary with the temperature, chemical characteristics, and type of organism to be destroyed. The amount required can best be determined by experience with the water being treated. Table 15.1 lists quantities of copper sulfate and chlorine required to kill algae.

Copper sulfate can be applied by various methods. For continuous feed, where the water passes through a conduit or channel, dry feed or solution feed directly to the flow can be utilized. Various methods of distribution from a boat, following a zigzag course over the surface area, are used for dosing the entire reservoir. Dragging bags filled with copper sulfate, scattering dry crystals over the surface with a blower mechanism, feeding dry crystals into the propeller wake of a power boat, and spraying a solution over the surface from a boat are several methods of distribution.

Table 15.1. Concentration Ranges for Copper Sulfate and Chlorine Required to Kill Algae[4]

Organisms	$CuSO_4$ ppm	Cl_2 ppm
Diatoms	0.1 to 0.5	0.5 to 2.0
Green	2.0 to 10.5	0.3 to 1.5
Blue-Green	0.1 to 0.5	0.5 to 1.1
Iron Bacteria	0.3 to 0.5	0.5
Sulfur Bacteria	5.0	0.5
Protozoa	0.1 to 2.0	0.3 to 1.0

Generally chlorine is used for algae control only when it can be added at the inlet to the reservoir. In certain instances chlorine has been added at 20 feet depths with satisfactory results.

Activated carbon is used to prevent the passage of sunlight into the water. This prevents the growth of algae, as sunlight is necessary for their reproduction. This treatment is termed "blackout." Methods of distribution are similar to copper sulfate treatment. Two tenths to 0.5 pound of powdered activated carbon per 1,000 square feet of surface area is generally used. Treatment is only practiced on sunny days.

15.4 CORRECTIVE TREATMENT

The most common practices of corrective treatment include one or a combination of the following treatments:[5]

1. Aeration
2. Coagulation
3. Chlorination
4. Chlorine Dioxide
5. Ozone
6. Potassium Permanganate
7. Activated Carbon

15.4.1 Aeration

Waters which are deficient in oxygen to the extent the taste is affected, can be improved by aeration. Hydrogen sulfide and free carbon dioxide are also reduced by aeration. Very little, if any, improvement is attained by aeration of waters with tastes and odors caused by microorganisms.

Where treatment requirements are seasonal, it is usually less expensive to use activated carbon or other corrective measures instead of aeration. However, where serious troubles are encountered, it is frequently cheaper to use aeration in conjunction with other methods. Aeration should be given careful study to determine its applicability to specific waters.

15.4.2 Coagulation

Good coagulation materially improves all other corrective treatments that are used in connection with taste and odor control. The economics of a single treatment method should be compared with methods employing combinations. Waters requiring taste and odor treatment usually contain other undesirable materials which make it necessary to employ methods that also aid taste and odor removal.

15.4.3 Chlorination

Chlorination as a corrective measure is most effective when free available residuals are produced and maintained. Chlorine is a strong oxidizing agent and converts many odorous compounds to unobjectionable forms. Also, very few organisms are able to exist in the presence of free available chlorine. Where chlorination has failed to control organic growth in treatment processes and distribution systems, it has been shown that residuals were composed primarily of chloramines, or interfering substances had produced false residuals.

Chlorine is applied at various points throughout the treatment plant. Points of application for best results will vary with each treatment facility. Applying it along with other chemicals in the mixing basins can provide, throughout the plant, a free residual, which will prevent the growth of algae and bacteria within the basins and filters.

Maintaining a free available residual throughout the distribution system will prevent tuberculation and organic accumulation, which attribute tastes and odors. Where it is impractical to apply sufficient chlorine at the treatment plant to maintain a free available residual throughout the distribution system, rechlorination at a point in the distribution system can be utilized.

Chlorine can intensify odors as well as eliminate them. Phenols are converted to chlorophenols when free available chlorine is not present.

Combined residual chlorination, treatment with ammonia and chlorine, prevents the formation of the most intense tastes and odors. The bactericidal efficiency of combined chlorine is much less than that of free chlorine. Waters with high ammonia or nitrogenous content usually yield equivalent results with chlorine alone.

Supplemental treatment is frequently required with chlorine to produce satisfactory results. This should not cause the elimination of pre-chlorination. More effective disinfection, improved coagulation, organic growth control, and other benefits are derived from chlorination.

15.4.4 Chlorine Dioxide

Chlorine dioxide is usually produced at the water plant by the reaction of sodium chlorite with chlorine solutions. Small installations without liquid chlorine generate chlorine dioxide with sodium chlorite, hypochlorite, and a mineral acid. It has high oxidizing power and is used primarily for control of phenolic type odors. It also serves as a disinfectant and permits carrying a residual throughout the distribution system. There is no simple method of determining chlorine dioxide residuals.

Pretreatment or post-treatment is satisfactory. Treatment prior to the filters is considered best. Post-treatment is used more frequently due to higher costs involved in pretreatment. Color removal and precipitation of iron and manganese are other benefits derived from the use of chlorine dioxide.

15.4.5 Ozone

Ozone is used as a disinfectant and to control taste and odor. The material does not form objectionable tastes and odors by reaction with phenolic compounds. Color removal and precipitation of iron and manganese are obtained by using ozone. Orthotolidine and iodometric methods can be used to measure residuals. Measurement is also adaptable to photometric procedure.

Ozone production requires equipment to clean, compress, cool, and dehumidify the air. The cleaned air is then passed through an electrical discharge apparatus to produce ozone. The yield is approximately 1 percent ozone. The large capital outlay for this equipment is the reason ozone has not been more widely accepted.

The practice of maintaining a residual throughout the distribution system has also limited its acceptance in the U.S.A.

15.4.6 Potassium Permanganate

Potassium permanganate has been found to be an effective, efficient and economical method of controlling taste and odor problems. Humphrey and Eikleberry[5] have evaluated the effectiveness of potassium permanganate in removing many common industrial and agricultural wastes and found that the quantity of potassium permanganate required to render the sample nonodorous is nearly a linear function of the concentration of the organic substance.

To use potassium permanganate effectively a minimum of one to one and one-half hours of contact time is required between the rapid mix and about three quarters of the way through the sedimentation basins. The water must be filtered because potassium permanganate is reduced to insoluble manganese oxide hydrates and all of the precipitate does not settle before reaching the filters. Control is possible by simple visual observation, because potassium permanganate oxidizes organic compounds and loses it characteristic pink color. By observing the sedimentation basin and keeping a pink color about one-half to three-quarters of the way through the basin will provide control.

Generally, the potassium permanganate is added at the low lift pumps or in the rapid mix. However, the point of application is dependent upon the characteristics of the water and the plant. The reaction rate of potassium permanganate is influenced by pH, and it is best to choose the point of application which will give optimum results.

Jar tests are used to evaluate the effectiveness of potassium permanganate and it is imperative that the tests simulate plant operation. The raw water should not be exposed to chlorination because chloro derivatives of organic compounds are more odorous, more difficult to oxidize, and more difficult to remove than the original compound.

In some instances it is more effective and economical to use potassium permanganate in conjunction with activated carbon. The potassium permanganate should be added first and allowed to be consumed before adding the activated carbon to obtain maximum taste and odor removal.

Potassium permanganate is available in bulk or in drums. Dry feeders are recommended for plants pumping over 1 mgd.

15.4.7 Activated Carbon

Taste and odor are removed with activated carbon by the process called adsorption. Adsorption is an adhesion which takes place at the surface of the activated carbon when taste and odor material make contact with the surface. Activated carbon can be produced by a wide variety of substances, but wood is the primary raw material. The raw material is charred at a temperature below 500° C and then activated by slow burning about 800° C. Activating agents include air, steam, chlorides, sulfates or phosphates, boric, nitric, phosphoric or sulfuric acid. Comparisons of adsorptive capacity are made by determining the "phenol value." The phenol value is defined as the mg/ℓ of activated carbon required to reduce 100 ppb of phenol by 90 percent. This value does not necessarily reflect the efficiency of the activated carbon in removing specific tastes and odors. The adsorptive capacity of ground activated carbon can easily be understood when it is recognized that one cu ft of the substance is estimated to have a surface area of about 300,000,000 square feet.

The fineness and porous character of activated carbon permits the carbon particles to disperse throughout the water and to remain in suspension for long periods of time. This provides sufficient time for the odorous substances to come in contact with the activated surfaces, which then hold these impurities on the surface. The carbon particles are then removed by coagulation, sedimentation, or filtration. Activated carbon removes the offensive material and does not convert it to a less offensive substance, nor does it provide a masking effect.

Activated carbon should be stored in a dry, fire-proof room or building. It burns without a flame when undisturbed, and if it is stored in a fire-proof building, the safest action is to allow it to smolder until it burns itself out. If it is necessary to extinguish the fire, a mist of water should be used, or, preferably, a carbon dioxide extinguisher. Keeping the material dry is necessary for proper operation of dry feeders.

Dry feed equipment should be designed to (a) prevent arching the hopper, (b) feed accurately and adjust easily, (c) wet carbon rapidly and carry it to the point of application, and (d) have sufficient capacity to handle emergency dosages. Dust collecting systems should also be provided to prevent escape of fine particles.

When activated carbon became available in bulk car load lots, many large plants converted to a slurry feeding method. The activated carbon is delivered to the plant in hopper bottom cars which are emptied into underground tanks. Here it is slurried with water and used as it is needed. Rotodip feeders, displacement pumps, and other proportional devices are used to pump the slurry. As simple as this procedure may sound, many difficulties were encountered before it became practical and efficient.

Where large dosages of carbon are required, it is recommended the carbon be applied appreciably ahead of the filters. When applied on the filters, some carbon may pass the filter, and the head losses will be tremendous. In a normal water treatment plant, carbon can be added to the raw water, mixing basin, settling basin, or onto the filters.

Exceptionally good results are often reported with very small dosages when added directly to the filters. The greatest efficiency from the carbon for taste and odor removal is usually reported for this type operation. Where odors are at low concentrations or appear infrequently, application directly to the filter is usually recommended.

Adsorption is most efficient with activated carbon when the pH is less than 9.0, because alkalies sometimes react with odorous substances to form salts, which are difficult to adsorb. Carbon should not be mixed with lime or soda ash, although it is permissible to feed them at the same point of application.

Chlorine compounds and carbon should never be mixed in a dry state. The reaction can be sufficiently violent to cause an explosion. This reaction does not occur in water; therefore, it is possible to apply both at a common point, but this is not recommended. Separation of application by fifteen minutes is preferable, because carbon adsorbs the chlorine and prevents the efficient use of both chemicals. Carbon is usually added prior to chlorination, because chlorinated compounds are more difficult to adsorb with activated carbon. Carbon also adsorbs organic compounds, which produce a chlorine demand, and thus reduces the chlorine required. The savings do not pay for the carbon, but, where carbon is required, the savings are substantial. Plant designs should provide means for flexible operation.

To determine the dosage of activated carbon, it is first necessary to find the threshold odor number which will produce a palatable water. This can best be accomplished by a taste and odor panel consisting of at least four individuals. Samples of raw water are chlorinated and filtered to insure a safe water for tasting. The water is then dechlorinated to prevent interference by chlorinous odors. This water is diluted with odor-free water, until a water acceptable to all four tasters is obtained. This is considered the palatable limit. A threshold odor test is then run on the dilution termed the "palatable limit." The threshold number will represent the value which must be attained in practice.

After determining the palatable limit, samples should be collected prior to the point of carbon application. To four samples, add various dosages of activated carbon; then treat each sample exactly as the water is treated in the treatment plant. Contact times should also parallel the plant detention times. Filter these samples and determine their threshold number. Also, determine the raw water threshold number. Plot these results on arithmetical graph paper as shown in Figure 15.1. From this curve, it should be possible to determine the activated carbon dosage required to obtain the desired threshold number.

Actual plant operation with activated carbon is superior to small scale laboratory tests. Therefore, it is recommended that one half the laboratory dosage be used on the first plant test. This usually gives the desired results.

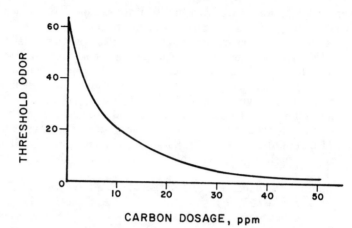

Figure 15.1. Determination of carbon dosage to control odors.

16. Water Softening

16.1 HARDNESS

16.1.1 Definition

Hardness generally refers to the sum of the calcium and magnesium ions in water expressed in milliequivalents per liter or more commonly as mg/ℓ of equivalent $CaCO_3$. Hardness in water is caused by the divalent metallic ions such as Ca^{++}, Mg^{++}, Fe^{++}, Mn^{++}, Ca^{++}, Ba^{++}, Zn^{++}, Pb^{++}, etc. All of these ions combine with ordinary soaps to form insoluble Ca-Mg soaps which are ineffective for cleansing purposes. Very little cleaning occurs until all of the divalent ions have combined with soap; therefore, soap is deactivated quantitatively and much more soap is required when hard waters are used for cleansing purposes.

16.1.2 Occurrence of Hard Waters

Divalent ions, principally calcium and magnesium, are dissolved from limestone, dolomite, and other minerals when water comes in contact. Groundwaters generally contain much greater concentrations of hardness than that found in surface waters. Hard waters are found throughout the world in areas where waters contact limestone or other calcium and magnesium bearing strata during recharge of aquifers. Details about hard water classification schemes and chemistry are presented in Section 9.4.2.

16.1.3 Types of Hardness

Sodium, calcium, and magnesium are the major ions found in water, and these cations are normally associated with the bicarbonate (HCO_3^-), chloride (Cl^-), sulfate ($SO_4^=$), and nitrate (NO_3^-) anions. The salts produced by the combination of the above anions and cations are classified as "temporary" or "carbonate" [i.e., $Ca(HCO_3)_2$ and $Mg(HCO_3)_2$] hardness and as "permanent" or "non-carbonate" hardness [i.e., $CaCl_2$, $MgCl_2$, $CaSO_4$, and $MgSO_4$]. Carbonate hardness can be removed by boiling the water which converts the HCO_3^- to $CO_3^=$ and finally to OH^-.

16.1.4 Disadvantages of Hardness

Increased soap consumption in hard waters is probably the most severe disadvantage of hardness. Much of this disadvantage has been eliminated by

239

the increased use of synthetic detergents, but where soaps are used the increased cost is significant. Hardness in water can clog skin pores, discolor porcelain, stain fabrics, and toughen and discolor vegetables. Process waters containing hardness when used in textile, paper, canning, and other industries can cause considerable deterioration in product quality. Boiler feed water containing hardness can cause scale buildup resulting in great losses in heat transfer. Boiler feed waters are invariably treated to remove excess hardness.

16.1.5 Advantages of Hardness

Deposition of a hardness film on pipe and other metal surfaces protects against corrosion. The presence of hardness in irrigation waters reduces the sodium ratio and generally improves production.

16.2 WATER SOFTENING

There are two basic methods of removing hardness or softening a water: 1) the precipitation method, and 2) ion exchange or the zeolite method. In general, the precipitation method is less expensive, but there are situations where ion exchange is advantageous and less expensive. Combinations of the two methods offer advantages in certain cases.

16.2.1 Precipitation Method

The lime-soda process is the principal process used to soften water; however, other chemicals are occasionally employed, i.e. in the hot processes used to treat boiler feed water. The basic reactions involved in the lime-soda process are presented below. The underlined compounds indicate precipitation from solution.

Removal of calcium and magnesium carbonate hardness:

$$Ca(HCO_3)_2 + Ca(OH)_2 \rightarrow \underline{2CaCO_3} + 2H_2O \quad \ldots \ldots \ldots \ldots (16.1)$$

$$Mg(HCO_3)_2 + Ca(OH)_2 \rightarrow \underline{Mg(OH)_2} + Ca(HCO_3)_2 \quad \ldots \ldots (16.2)$$

$Ca(HCO_3)_2$ produced is removed as shown for natural compound.

Removal of calcium non-carbonate hardness:

$$CaCl_2 + Na_2CO_3 \rightarrow \underline{CaCO_3} + 2NaCl \quad \ldots \ldots \ldots \ldots \ldots (16.3)$$

Removal of magnesium non-carbonate hardness:

$$MgCl_2 + Ca(OH)_2 \rightarrow \underline{Mg(OH)_2} + CaCl_2 \quad \ldots \ldots \ldots \ldots (16.4)$$

$$CaCl_2 + Na_2CO_3 \rightarrow \underline{CaCO_3} + 2NaCl \quad \ldots \ldots \ldots \ldots \ldots (16.5)$$

Removal of free carbon dioxide:

$$CO_2 + Ca(OH)_2 \rightarrow CaCO_3 + H_2O \quad \ldots \ldots \ldots \ldots \ldots (16.6)$$

16.2.2 Chemical Requirements

16.2.2.1 Stoichiometric Method

The total amount of lime and soda ash required to soften a water can be estimated from the above expressions. A simple procedure is to use the following expressions which is based on the above stoichiometric relationships with all concentrations expressed in terms of calcium carbonate.

$$\text{Lime, mg/}\ell \text{ as } CaCO_3 = Mg + \text{Bicarbonate alkalinity} + CO_2 \quad \ldots (16.7)$$

$$\text{Soda ash, mg/}\ell \text{ as } CaCO_3 = Ca + Mg - \text{Bicarbonate alkalinity} \quad \ldots (16.8)$$

$$\text{Lime, mg/}\ell \text{ as } CaO = 0.56 \, (Mg + \text{Bicarbonate alkalinity}$$
$$+ CO_2) \ldots \ldots \ldots \ldots \ldots \ldots \ldots (16.9)$$

$$\text{Soda ash, mg/}\ell \text{ as } Na_2CO_3 = 1.06 \, (Ca + Mg$$
$$- \text{Bicarbonate alkalinity}) \ldots \ldots \ldots \ldots (16.10)$$

$$\text{Lime, lbs CaO/M.G.} = 4.66 \, (Mg + \text{Bicarbonate alkalinity}$$
$$+ CO_2) \ldots \ldots \ldots \ldots \ldots \ldots (16.11)$$

$$\text{Soda ash, lbs } Na_2CO_3/\text{M.G.} = 8.83 \, (Ca + Mg$$
$$- \text{Bicarbonate alkalinity}) \ldots \ldots \ldots \ldots (16.12)$$

When the bicarbonate alkalinity is greater than the sum of the Ca and Mg, the above equations must be corrected by reducing the value used for bicarbonate alkalinity to an amount equal to the sum of the Ca and Mg. Soda ash will not be required in this case.

Example 16.1

Calculate the theoretical quantity of lime and soda ash required to treat a water with the following characteristics expressed as the ion or compound.

$$
\begin{aligned}
CO_2 &= 9 \text{ mg/}\ell \\
Ca &= 152 \text{ mg/}\ell \\
Mg &= 29 \text{ mg/}\ell \\
HCO_3^- &= 317 \text{ mg/}\ell \\
\text{pH value} &= 7.0
\end{aligned}
$$

Solution:

a.   Express concentrations in terms of $CaCO_3$:

$$CO_2 = 9 \text{ mg/}\ell \left(\frac{50}{22}\right) = 20 \text{ mg/}\ell \text{ as } CaCO_3$$

$$Ca = 152 \text{ mg/}\ell \left(\frac{50}{20}\right) = 380 \text{ mg/}\ell \text{ as } CaCO_3$$

$$Mg = 29 \text{ mg/}\ell \left(\frac{50}{12.15}\right) = 120 \text{ mg/}\ell \text{ as } CaCO_3$$

$$HCO_3^- = 317 \text{ mg/}\ell \left(\frac{50}{61}\right) = 260 \text{ mg/}\ell \text{ as } CaCO_3$$

b.   Calculate lime required:

Lime, mg/ℓ as $CaCO_3$ = Mg + Bicarbonate alkalinity + CO_2
= 120 + 260 + 20
= 400 mg/ℓ

c.   Calculate soda ash required:

$Na_2 CO_3$, mg/ℓ as $CaCO_3$ = Ca + Mg - Bicarbonate alkalinity
= 380 + 120 - 260
= 240 mg/ℓ

More lime is required than indicated by the stoichiometric relationships, and it is common practice to add 25 to 50 mg/ℓ of excess lime to force the reactions to completion. The "excess lime" represents hardness and has supersaturated the water. To prevent precipitation in the water distribution system, it is necessary to remove the "excess lime" after settling. Recarbonation with CO_2 gas is the usual method of removing "excess lime."

16.2.2.2 Caldwell-Lawrence Diagram

The major disadvantage to using the stoichiometric method of estimating softening chemical dosages is that the results with intermediate dosages of chemicals cannot be estimated. Using the Caldwell-Lawrence[1] water conditioning diagram shown in Figure 16.1, it is possible to calculate the characteristics of a water treated with various dosages of lime and soda ash.

Example 16.2

Using the Caldwell-Lawrence diagram calculate the characteristics of the water described below when treated with various dosages of lime and 60 mg/ℓ of soda ash. Plot the results showing residual Ca^{++} and Mg^{++}, pH value, and the alkalinity in the treated water as a function of the lime dosage.

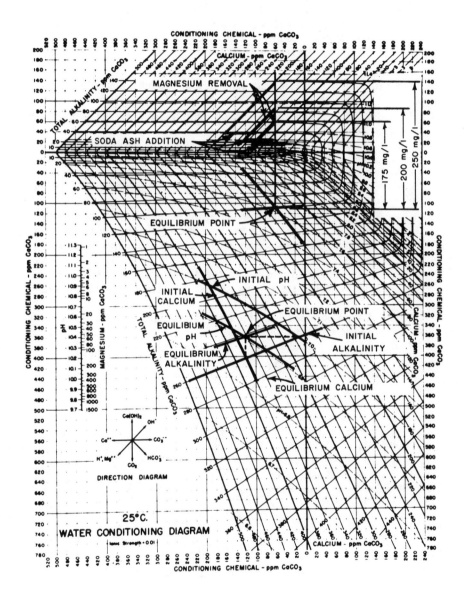

Figure 16.1. The Caldwell-Lawrence water conditioning diagram.[1]

Solution:

a.   Characteristics of untreated water:
     Calcium—160 mg/ℓ as $CaCO_3$
     Magnesium—100 mg/ℓ as $CaCO_3$
     Alkalinity—100 mg/ℓ as $CaCO_3$
     pH value—7.65

b.   Identify raw water characteristics on Caldwell-Lawrence diagram (Figure 16.1).

c.   Extend a line from the equilibrium point to the top of the diagram (Figure 16.1).

d.   Select various dosages of lime beginning at the equilibrium point and progress up the line extended in step c. At each dosage selected, the characteristics of the water can be read from the diagram. After reaching the minimum alkalinity line (zero on the ordinate showing quantities of conditioning chemicals), magnesium is precipitated and the quantity removed is determined by projecting a line to the left at a 45° angle a distance equal to the distance the dosage goes above zero or the minimum alkalinity line.

e.   From the point reached in step d project a horizontal line to the right equal to the 60 mg/ℓ of soda ash. This is the final condition of the water.

f.   Values obtained from Figure 16.1 are summarized in Table 16.1 and plotted on Figure 16.2.

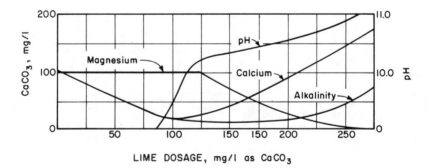
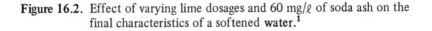

Figure 16.2. Effect of varying lime dosages and 60 mg/ℓ of soda ash on the final characteristics of a softened water.[1]

Table 16.1. Results Obtained from Caldwell-Lawrence Diagram for Varying Dosages of Lime and 60 mg/ℓ of Na_2CO_3

Ions and pH Value	Lime Dosage, mg/ℓ as $CaCO_3$			
	100	175	200	250
Ca^{++}	20	60	90	145
Mg^{++}	100	58	37	9
Alkalinity	20	17	25	40
pH	9.8	10.4	10.5	10.8

Example 16.3

Calculate the equilibrium point for a water not naturally equilibrated using the Caldwell-Lawrence diagram. The characteristics of the water are as follows:

Ca = 380 mg/ℓ as $CaCO_3$
Mg = 120 mg/ℓ as $CaCO_3$
Alkalinity = 260 mg/ℓ as $CaCO_3$
pH = 7.0

Solution:

a. Sketch lines on the Caldwell-Lawrence diagram (Figure 16.1) describing the natural water.
b. Project a vertical line upward from the intersection of the calcium and alkalinity lines, and project a horizontal line from the intersection of the alkalinity line and the pH line. The point of intersection for the two projections is the equilibrium point (Figure 16.1).
c. To calculate conditioning chemical dosages in waters unequilibrated, the same procedure outlined in Example 16.2 is followed beginning at the equilibrium point.

16.2.3 Variations of Lime-Soda Ash Process

16.2.3.1 Conventional Softening

The conventional municipal water softening plant consists of the addition of chemicals, mixing, settling, carbonation, and filtration. A flow diagram for a conventional municipal water softening plant is shown in Figure 16.3. Mixing and settling can be carried out in separate tanks or in "sludge blanket" vertical flow systems. In "sludge blanket" systems previously formed precipitate is suspended due to the upward velocity of the water, and clarification is aided by the upward filtration through the sludge, or

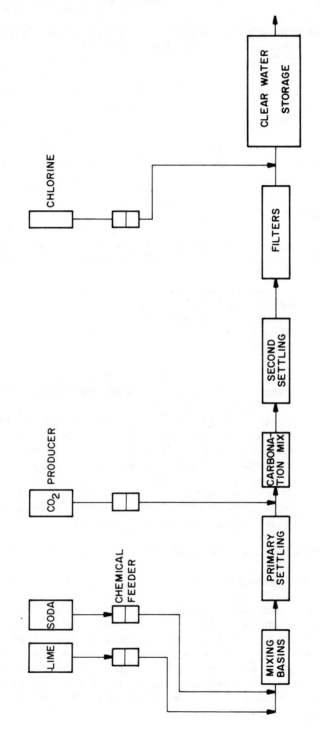

Figure 16.3. Flow diagram for lime-soda softening.

precipitate, blanket. Typical "sludge blanket" proprietary units are shown in Figures 16.4 and 16.5.

In the conventional process carbonate hardness is reduced to a level of 35-40 mg/ℓ and the total hardness is lowered to 50-100 mg/ℓ. Part of the non-carbonate hardness frequently is not removed because of the higher price of sodium carbonate. It is also desirable to have some hardness in municipal water supplies to improve taste and the rinsing properties.

Softened water must be stabilized with respect to $CaCO_3$ to prevent deposition in the distribution system. Recarbonation with CO_2 gas followed by sedimentation is the most common method employed to stabilize the softened water.

16.2.3.2 Excess Lime Softening

Excess lime softening is practiced in two ways: with recarbonation and without recarbonation. In both methods excess lime and soda ash are added to the water to be softened in an amount sufficient to increase the total

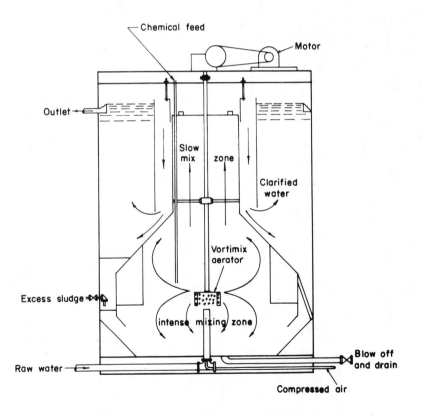

Figure 16.4. Aero-accelator. (Courtesy of Infilco Degremont.)

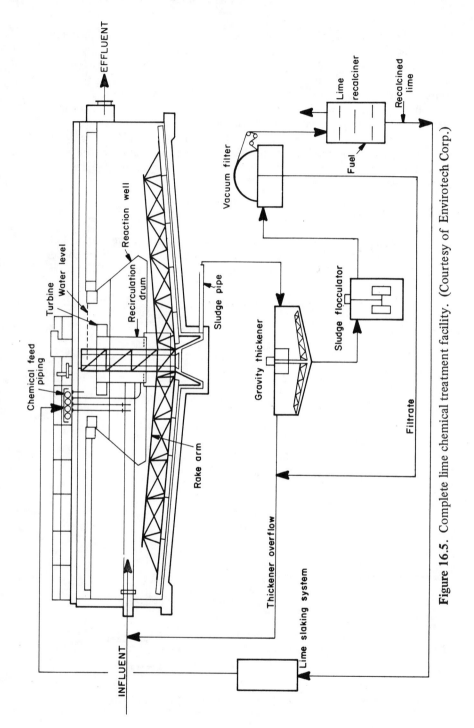

Figure 16.5. Complete lime chemical treatment facility. (Courtesy of Envirotech Corp.)

alkalinity of the treated water to a value of 35 mg/ℓ higher than the hardness. A high concentration of caustic alkalinity is present under these conditions. A water with a hardness of 30-50 mg/ℓ can be produced by this method. The caustic alkalinity is removed by recarbonating with CO_2 gas to produce an acceptable water. Excess lime softening without recarbonation does not produce an acceptable water because of the excess caustic alkalinity. However, this objection can be overcome as described under split treatment.

16.2.3.3 Split Treatment

A large portion of the raw water is treated with excess chemicals, and after preliminary sedimentation, raw water is added to neutralize the excess chemicals. Then the mixture is filtered and disinfected. Where split treatment is used, a hardness removal per unit of chemical is greater than for the conventional or excess treatment with recarbonation. The residual hardness is 50-100 mg/ℓ.

16.2.3.4 Lime and Coagulation

The use of coagulants in conjunction with softening chemicals can increase removals or shorten the settling times required to produce an acceptable product. The increase in efficiency or reduction in size of sedimentation basins must be evaluated economically to determine if the use of coagulants is justified.

16.2.3.5 Secondary Stirring

Excess chemicals are added as described above and following sedimentation the excess lime is partially recarbonated leaving the supersaturated $CaCO_3$. Coagulant or sludge from the softening operation is then added to the water and mixed and allowed to settle. The water is carbonated and filtered. Secondary stirring produces a more stable filtered water, causes less incrustation of filter sand grains, and decreases the load of material passing on to the filter.

16.2.3.6 Hot Processes

Most boiler feed water is preheated and softened before it is added to the boiler. Heating drives the reactions closer to completion and accelerates precipitation and aggregation. The solubility products of the precipitates are also reduced. Excess lime and recarbonation are not required. A water with a residual hardness of 30-35 mg/ℓ can be produced with a total reaction and settling time of approximately one hour. Silica filter media must not be used in the hot process to avoid solution of the silica and the production of analcite scale in the boiler. Anthracite coal is normally used as the filter media.

Hot phosphate softening is a variation frequently employed. NaH_2PO_4, $NaHPO_4$, and $NaOH$ or Na_3PO_4 are used instead of lime and soda ash or these compounds are used following lime and soda ash treatment. Residual hardness of approximately 5 mg/ℓ is produced. Sulfuric acid is usually added to prevent after-precipitation and to adjust the pH value.

16.2.3.7 Lime-Zeolite Process

Salt to recharge zeolite or other resins is less expensive than soda ash. Therefore, many plants use lime to remove the carbonate hardness followed by the zeolite process to remove the non-carbonate hardness. A flow sheet for this process is shown in Figure 16.6. Zero hardness can be produced with exchange resins, and a final water of any desired hardness can be produced by blending water from the lime treated step with zeolite treated water.

16.2.4 Disposal of Sludges

Sludge disposal constitutes one of the most difficult problems in water softening. Sludges cannot be discharged to a stream, and discharge to a sewage treatment plant can upset the operation because of the high alkalinity. If land is available, lagooning or application to farm land as a conditioner has proven satisfactory. Dewatering with centrifuges or vacuum filters is practiced in many locations. The most effective method is the recovery and reuse of the lime by calcining the $CaCO_3$ to CaO in a rotary kiln or multiple hearth furnace after dewatering. Accumulation of inert MgO can be a problem and eventually the recovered lime becomes ineffective. There are processes available to avoid Mg accumulation and they should be used where Mg is present in large quantities. CO_2 released from the calcining operation can be used to recarbonate.

16.2.5 Stabilization

After softening and in many natural waters, it is necessary to stabilize the treated water to eliminate caustic alkalinity and the after-precipitation of excess $CaCO_3$. Failure to stabilize softened waters results in incrusted pipes and enlarged sand grains in filters. Certain natural waters may be corrosive and require the addition of alkalinity. Methods available to stabilize waters are as follows:

a. Addition of mineral acids, H_2SO_4 or $NaHSO_4$ to convert carbonates into bicarbonates, hydroxides to sulfates and phosphates. Difficult to control and used infrequently.

b. Recarbonation with CO_2 gas which converts carbonates and hydroxides to bicarbonates.

c. Phosphate stabilization uses sodium hexametaphosphate $(NaPO_3)_6$, or sodium pyrophosphate $(Na_4P_2O_7)$ to inhibit the precipitation of $CaCO_3$. Dosages of 0.5 to 1 mg/ℓ are usually

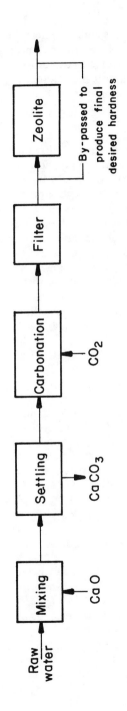

Figure 16.6. Flow diagram for the lime-zeolite process.

adequate to stabilize waters for long periods of time. Used in most small plants and in some larger plants because the process is easy to control and costs are approximately equal.

d.   Chelating agents are used to bind the metallic ions and prevent their deposition.

16.2.6 Recarbonation

Control of recarbonation is difficult because of the time required to obtain equilibrium. When true equilibrium is reached, after-deposition will still occur, and if recarbonation is sufficient to prevent after-deposition, the water becomes corrosive when true equilibrium is reached. Recarbonation is usually used to adjust the pH value between 9.0 and 9.6 leaving the water slightly supersaturated. (A pH value of 8.3 would indicate equilibrium.) If scaling occurs, polyphosphates can be used to control deposition.

A conventional plant will produce CO_2 in a combustion chamber which is followed by a scrubber, drier, blower, and recarbonation basin. Fuels may be gas, coke, oil, or a mixture of the three. Commercial CO_2 is available and used at some installations. Carbon dioxide is added to the water in the recarbonation basin through conventional perforated pipe diffusers or proprietary devices. Carbon dioxide is usually added prior to filtration.

16.3 ZEOLITE OR ION EXCHANGE PROCESS

16.3.1 Types of Mineral Zeolites

Mineral zeolites are insoluble double dilicates of sodium or potassium with aluminum or iron and water of hydration and are represented as follows: $Na_2O \cdot AlO_3 \cdot (SiO_2)_x \cdot (H_2O)_n$. Sodium and potassium are reversibly exchangeable for the alkaline earth group (Ba, Ca, Mg, Sr) ammonia, and some of the divalent metals such as iron and manganese. An example of the ion exchange reactions is shown in Equation 16.13.

$$Na_2Z + Ca^{++} \rightleftharpoons CaZ + 2Na^+ \quad \cdots \cdots \cdots \cdots \cdots \quad (16.13)$$

$$Na_2Z + Mg^{++} \rightleftharpoons MgZ + 2Na^+ \quad \cdots \cdots \cdots \cdots \cdots \quad (16.14)$$

Equations 16.13 and 16.14 favor the formation of CaZ and MgZ, but at high concentrations of Na^+ the reaction is reversed. Brine solutions are used to recharge the zeolite. Both the softening and the recharging reactions are rapid. A water with a hardness of essentially zero (1-5 mg/ℓ) can be produced by the zeolite process.

Greensand (glauconite) or natural zeolites are produced by treatment with heat and sodium hydroxide. Greensand has an exchange capacity of 5,700 to 6,900 grams of hardness per m^3. Synthetic zeolites are manufactured by mixing solutions of Na_2SiO_3 and alum or $NaAlO_2$ which form a gel of $Al(OH)_3$ and H_2SiO_3. The gel is dried and ground to size. Synthetic zeolite has a density of approximately 800 kg/m^3 which is approximately one-half that of natural greensands. Synthetic zeolite has a softening capacity of 14,000 to 17,000 grams per m^3.

Natural greensands require 3.5 to 7.0 grams of salt per gram of hardness removed for regeneration and a regeneration time of 20 to 30 minutes. Salt requirements for synthetic zeolite range from 2.5 to 3.5 grams per gram of hardness removed with regeneration times of 30 minutes or more. Synthetic zeolites are less durable than natural ones and CO_2 concentrations above 15 mg/ℓ will disintegrate the synthetic zeolite.

16.3.2 Type of Systems

A zeolite softening plant can consist of gravity or pressure vessels, similar to activated carbon or sand filters, with 76 to 200 cm of zeolite with an effective size of 0.25-0.30 mm for natural materials and 0.42 mm for synthetic zeolites. The direction of flow can be upward or downward at flow rates ranging from 0.16 to 0.32 m^3/m^2-min. Upflow units are usually limited to filtered waters such as in the lime-zeolite process. Auxiliary equipment consists of brine-production tanks, meters, and pumps.

The softening operation involves passing water through the bed of zeolite until the zeolite becomes saturated. The bed is then backwashed to remove suspended matter, brine (5-10 percent solution) is added and held in the bed for approximately 30 minutes or passed through the bed slowly, and the brine is withdrawn and the bed is rinsed to remove excess salt. Following the rinse, softening is resumed. The entire operation is frequently automated.

16.3.3 Advantages and Disadvantages

Advantages of zeolite softening are the compactness of the system, simple to operate, can be automated, no lime sludge problem, and double pumping is eliminated. The disadvantages include a brine disposal problem and the following situations where zeolite is inapplicable:

1. Cannot be used on waters containing turbidity, oil, or H_2S which coat the zeolite and make it inactive.
2. Cannot be used on waters containing more than 2 mg/ℓ of Fe^{++} or Mn^{++} because both will oxidize and precipitate on grains.
3. Cannot be used on waters with hardness concentrations above 800 mg/ℓ because of the low yield per unit of zeolite and the frequent regeneration required.
4. Cannot be used on waters with high sodium concentrations because of low efficiencies and the addition of more sodium which would make water objectionable in many applications.

16.3.4 Synthetic Ion Exchange Resins

Synthetic resins are used in most ion exchange applications because of the greatly increased capacity and the wide selection of special application resins. Other than the need for a low turbidity water and the absence of oil, the disadvantages listed for zeolite softening can be eliminated by the proper design and selection of a synthetic resin. The basic design considerations and advantages listed for zeolite softening are applicable to synthetic resin systems.

Styrene base cation resins with sulfonic acid functional groups are widely used for softening water. These resins have a high capacity ($>$ 100 kg of hardness/m^3 of resin), excellent stability, and are selective for Ca^{++} and Mg^{++}. Salt requirements vary from 46 to 155 kg of NaCl per m^3 of resin depending upon the percentage of the exchangeable spaces utilized.

The softening exchange reactions normally utilized in water softening are as follows:

$$2RNa + Ca^{++} \rightleftharpoons R_2Ca + 2Na^+ \quad \ldots \ldots \ldots \ldots \quad (16.15)$$

$$2RNa + Mg^{++} \rightleftharpoons R_2Mg + 2Na^3 \quad \ldots \ldots \ldots \ldots \quad (16.16)$$

These reactions are the same as those shown in Equations 16.3 and 16.4 except for the substitutions of the synthetic resin.

There are many synthetic resins available and when the use of resins is to be evaluated, it is advisable to contact the manufacturers and consider many alternatives. The field is continuously improving and expanding and the most efficient and economical design can be obtained only if the most current resins are employed.

17. Iron and Manganese Removal

17.1 EFFECT AND OCCURRENCE OF IRON AND MANGANESE

Iron and manganese in water precipitate and produce an undesirable turbid yellow-brown to black water which stains laundry and plumbing fixtures. Iron and manganese also support the growth of microorganisms in distribution systems. These growths can accumulate and reduce the effective area of the pipe and clog meters and valves. When the growth sloughs off, taste and odor problems occur. Several milligrams per liter of iron and manganese will impart a "metallic" taste.

Groundwater supplies require iron and manganese removal far more frequently than surface waters. Only in unusual situations, such as a surface supply taken from the hypolimnion of deep lakes or reservoirs, is iron and manganese removal required with surface supplies.

17.2 METHODS FOR REMOVING IRON AND MANGANESE

There are three general methods normally employed to remove iron and manganese: 1) precipitation and filtration, 2) ion exchange, and 3) stabilization with dispersing agents. The third method is usually limited to systems without filtration and waters containing less than 1 mg/ℓ of iron and manganese.

Precipitation and filtration to remove iron and manganese are performed in one of the three systems illustrated in Figure 17.1. Aeration followed by sedimentation (short contact period) and filtration is the most common method employed (Figure 17.1a). Chlorine is frequently added following aeration to aid the oxidation. Lime is added to adjust the pH value of waters too acid for rapid oxidation with aeration. Chlorine, chlorine dioxide, and potassium permanganate are used most frequently with waters containing manganese because manganese is more difficult to oxidize (Figure 17.1b). Potassium permanganate will rapidly oxidize manganous iron to manganese dioxide over a wide range of pH values. However, the pH value is normally maintained between 7.5 and 8.0 because of the difficulty associated with the manganese dioxide floc formation and filterability at higher pH values. In some instances chlorine is added before the permanganate to oxidize the more easily removed iron. Chlorine is used because it is generally less expensive than permanganate or chlorine dioxide.

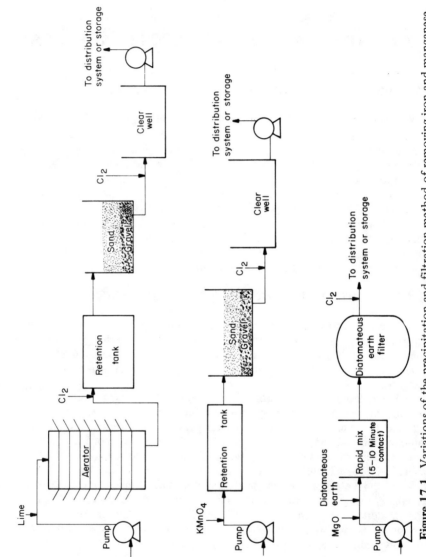

Figure 17.1. Variations of the precipitation and filtration method of removing iron and manganese.

Calcined magnesite has been employed to oxidize iron and manganese (Figure 17.1c). The magnesite and diatomaceous earth are added to the rapid-mix tank with a contact time of 5-10 minutes, and then the mixture is filtered through a diatomaceous earth filter. Neither precipitates nor bacterial growth accumulate in this process because the filter media is discarded at the end of each cycle.

Ion exchange resins can effectively remove iron and manganese in the absence of dissolved oxygen where the concentration is less than 0.5 mg/ℓ. With higher concentrations the oxide precipitates of these metals can cause the resin to become saturated and ineffective. Because of the coating problem, regular ion exchange processes are not normally used to remove iron and manganese. Modifications of the ion exchange or zeolite process have been successful.

· Two modifications have been used and the processes are an intermittent regeneration process and a continuous process using manganese treated greensand (zeolite) as both the oxidant and the filter media. In the intermittent regeneration process (Figure 17.2) water passes through the mineral bed and the iron and manganese are oxidized as described in Equations 17.1, 17.2, 17.3, and 17.4.

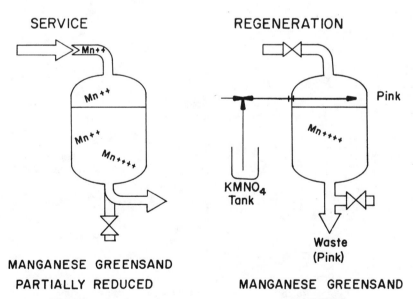

Figure 17.2. Intermittent regeneration of manganese greensand. (Courtesy of Carus Chemical Co., Inc.)

Exchange:

$$Na_2Z + Mn^{++} \rightarrow MnZ + 2Na^+ \quad \ldots \ldots \ldots \ldots \quad (17.1)$$

Generation:

$$MnZ + KMnO_4 \rightarrow MnO_2 \cdot Z + K^+ \quad \ldots \ldots \ldots \ldots \quad (17.2)$$

Degeneration:

$$MnO_2 \cdot Z + Fe^{++} \text{ or } Mn^{++} \rightarrow Mn_2O_3 \cdot Z$$
$$+ Fe^{+++}, Mn^{+++}, \text{ or } Mn^{4+}. \quad \ldots \ldots \ldots \ldots \quad (17.3)$$

Regeneration:

$$Mn_2O_3 \cdot Z + KMnO_4 \rightarrow MnO_2 \cdot Z \quad \ldots \ldots \ldots \ldots \quad (17.4)$$

The principal disadvantages of the intermittent regeneration process are economic and manganese leakage near the end of the cycle. Large quantities of excess potassium permanganate are required to regenerate the manganese treated greensand, and large volumes of water are required to rinse the excess $KMnO_4$ from the regenerated greensand.

In the continuous process potassium permanganate is added continuously ahead of the filter containing anthracite coal filter media and manganese treated greensand (Figure 17.3). The chemistry of the reactions is the same as presented in Equations 17.1 through 17.4. $KMnO_4$ in solution

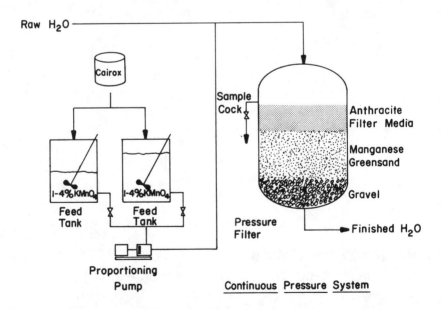

Figure 17.3. Processing utilizing a continuous feed of $KMnO_4$ solution ahead of a filter containing anthracite filter media and manganese treated greensand. (Courtesy of Carus Chemical Co., Inc.)

oxidizes the iron and manganese more efficiently than the manganese treated greensand. The manganese treated greensand serves as a reserve oxidant and as a filter. If the dosage of $KMnO_4$ is low with respect to the concentration of iron and manganese, the greensand will remove the portion unoxidized. When the excess of $KMnO_4$ is added, the greensand will be regenerated (Equation 17.4). The application of $KMnO_4$ is controlled visually. The application of $KMnO_4$ should be adjusted such that a slight increase in dosage will produce a pink color in the filter effluent water. The anthracite filter media above the manganese treated greensand increases filter runs by removing the bulk of the insoluble precipitates. The continuous process reduces the down time by two-thirds by eliminating the time required for regeneration and washing excess $KMnO_4$ from the filter.

Potassium permanganate can also be used in conventional water treatment plants to oxidize iron and manganese and to control tastes and odors. A typical system is shown in Figure 17.4.

Sodium hexametaphosphates are used to stabilize iron and manganese when the concentration of both does not exceed 1 mg/ℓ. Approximately 5 mg of $(NaPO_3)_6$ is required per mg of iron plus manganese. When the water is heated, the polyphosphate reverts to orthophosphate and no longer disperses the iron and manganese. Polyphosphates must be added before the iron and manganese are oxidized because the polyphosphates do not effectively stabilize the precipitated ferric hydroxide.

17.3 EQUIPMENT EMPLOYED

Aeration, detention, and filtration are the most common methods of removing iron and manganese. In addition to oxidizing iron and manganese, aeration reduces the CO_2 content of most groundwaters increasing the pH value. Methane and hydrogen sulfide are removed by aeration; methane readily and H_2S partially.

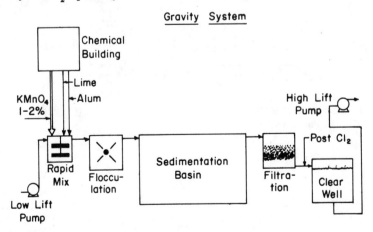

Figure 17.4. Conventional gravity system utilizing potassium permanganate.

Aeration is normally performed in a contact aerator (Figure 17.5) consisting of trays containing coke, limestone, wooden slats, etc. A short detention time must be provided to allow the oxidation to go to completion although most of the iron and manganese precipitates do not settle but are trapped on the filter surface. Detention should be provided even when strong oxidants such as chlorine and potassium permanganate are used. The proper detention time required is site specific and should be determined with jar tests.

Filter sand used to remove oxidized iron and manganese are usually coarser (effective size ≈ 0.6 mm) than those used with coagulated waters. Pressure filters frequently encounter mud ball problems, or cementing of sand grains by deposition of iron, manganese and calcium. A well designed backwash system can aid in preventing this problem.

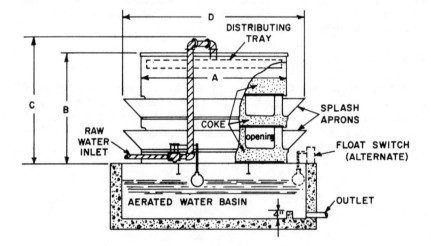

Figure 17.5. A cross sectional view of a coke tray aerator. (Courtesy of Permutit.)

References

Chapter 1

1. Renn, E. C. A remarkable compound. Span., U.S. Embassy in India, May 1967.

2. Müller-Neuhaus, G. Wasserwirtschaftliche probleme in der modernen industriegesellschaft. Nuro-päisches Abwassersymposium, München, 1969, Berichte der Abwassertechnischen Vereinigung.

3. United States Water Resources Council. The nation's water resources. Washington, D.C., U.S. Government Printing Office, 1968.

Chapter 2

1. Indian Standard Institution. I.S. code of basic requirements for water supply, drainage and sanitation. I.S. 1172, 1957.

2. Wasserversorgung, Lehrbrief 1, TH. Dresden, VEB Verlag, Technik, Berlin, 1961.

3. Bohnke, B. Volkswirtschaftlicher aufwand für die wasserversorgung sowie für die beseitigung flüssinger and fester Abfallstoffe Luft, Wasser and Betrieb, Vol. 15, No. 8, August 1971.

4. McJunkin, F. E. Population forecasting. Journal of the Sanitary Engineering Division, ASCE, August 1964.

Chapter 3

1. Wassergute, Lehrbrief, TH. Dresden, 1961.

Chapter 4

1. Ahmad Nazir. Tube wells construction and maintenance. Scientific Research Stores, Lahore, 1969.

261

2. Dupuit, J. Études theoriques et pratiques sur la mouvement des eaux dans les canaux découverts et à travers les terrains perméables. 2nd ed., Dunod, Paris, 1863.

3. Sichardt, W. Das Fassungsvermögen von Rohrbrunnen und seine Bedeutung für die Grundwasserabsenkung, inbesondere für grossere Absenkstiefen. J. Springen, Berlin. 1928.

4. Theis, C. V. The relation between the lowering of piezometric surface and the rate and duration of discharge of a well using ground-water storage. Trans. Amer. Geophysical Union, Vol. 16, 1935.

5. Jacob, C. E. Flow of ground water. Engineering Hydraulics, Rouse, H., editor, John Wiley and Sons, New York, 1950.

6. Wenzel, L. K. Methods for determining permeability of water-bearing materials with special reference to discharging-well method. U.S. Geological Survey Water Supply Paper 887, Washington, D.C., 1942.

7. Wasserversorgung, 1. Lehrbrief. TH. Dresden, VEB Verlag Technik, Berlin, 1961.

8. Busch, K. F. Wasserversorgung für Stadt and Landwirtschaft. B. G. Teubner Verlagsgesellschaft, Leipzig, 1956.

9. Casagrande, Arthur. Classification and identification of soils. Trans., Am. Soc. of Civil Engineers, 113; 901. 1948.

10. Brix-Heyd-Gerlach. Die Wasserversorgung. Verlag von R. Oldenbourg. Munchen, 1952.

Chapter 5

1. Steel, E. W. Water supply and sewerage. McGraw-Hill, 1960.

2. Busch, K. F. Wasserversorgung für Stadt und Landwirtschaft, B. G. Teubner Verlagsgesellschaft, Leipzig, 1956.

Chapter 7

1. Davis, C. V. Handbook of applied hydraulics. McGraw-Hill Book Co.

Chapter 8

1. Wasserversorgung, 6. Lehrbrief für das Fernstudium, 1964, TH. Dresden, VEB Verlag, Technik, Berlin.

2. Liebhold. Vereinfachte Berechnung der Druckhohen in Rohrleitungen, Gesundheitsingenieur, 81, Jg. 1960, H. 12.

3. Schulz, H. Tabellenbuch für die Berechnung von Rohrleitungen and Kanälen in Siedlungswasserbau. VEB Verlag, Technik, Berlin, 1959.

Chapter 9

1.. Sawyer, C. N., and P. L. McCarty. Chemistry for sanitary engineers. McGraw-Hill, 1967.

2. Standard methods for the examination of water and waste water. American Public Health Association, 14th edition, 1975.

Chapter 11

1. Camp, T. R. Studies of sedimentation basin design. Sewage Ind. Wastes, Vol. 25, No. 1 (January 1953).

2. Slechta, A. F., and W. R. Conley. Recent experiences in plant-scale application of the settling tube concept. Journal Water Pollution Control Federation, Vol. 43, p. 1724, 1971.

Chapter 12

1. Sen, A. K., and K. R. Bulusu. Effectiveness of Nirmali seed as coagulant and coagulant aid. Environmental Health, p. 233, October 1962.

2. Rushtan, J. H. Mixing of liquids in chemical engineering. Ind. Engineering Chemistry, Vol. 44, 1952.

3. Camp, T. R. Flocculation and flocculation basins. Trans. ASCE, 120, 1. (1955).

Chapter 13

1. Camp, T. R. Hydraulics of filtration. ASCE, p. 1, August 1964.

2. Iwasaki, T. Some notes on sand filtration. Journal American Water Works Association, Vol. 29, p. 1591, 1937.

3. Ives, K. J., and I. Scholji. Research on variables effecting filtration. Journal Sanitary Engineering Division, ASCE, Vol. 91, p. 1-18, 1965.

4. Ives, K. J. Rational design of filters. Proceedings, Institution of Civil Engineers, London, England, Vol. 16, p. 189, 1960.

5.   Fox, D. M., and J. L. Cleasby. Experimental evaluation of sand filtration theory. Journal Sanitary Engineering Division, ASCE, Vol. 92, p. 61, 1966.

6.   Foust, A.B., et al. Principles of unit operations. John Wiley, 1960, 475 p.

7.   Fair, G. M., and J. C. Geyer. Water supply and waste water disposal. John Wiley and Sons, 1954.

Chapter 14

1.   Fair, G. M., and J. C. Geyer. Water supply and waste water disposal. John Wiley and Sons, 1954.

2.   Sawyer, C. M. Tracer studies on a model chlorine contact tank. M.S. thesis, Virginia Polytechnic Institute Library, Blacksburg, Va., 1967.

3.   Marske, Donald M., and Jerry D. Boyle. Chlorine contact chamber design—a field evaluation. Water and Sewage Works 120(1):70-77, 1973.

4.   Wolf, D., and W. Resnick. Residence time distribution in real systems. Industrial and Engineering Chemistry Fundamentals 2(4):287-293, 1963.

5.   Rebhum, M., and Y. Argaman. Evaluations of the hydraulic efficiency of sedimentation basins. Proceedings of the ASCE Sanitary Engineering Division 91(SAS):37-45, 1965.

6.   Stephenson, R. L., and J. R. Lauderbaugh. Baffling chlorine contact tanks. Water and Sewage Works, Reference Number 1971, R-100-103, 1971.

Chapter 15

1.   Middlebrooks, E. J. Taste and odor control. Water and Sewage Works, Reference Number 1965.

2.   Silvey, J. K. G., J. C. Russel, D. R. Redden, and W. C. McCormick. Actinomycetes and common tastes and odors. Jour. AWWA, 42:11; 1018, 1950.

3.   Hoak, R. D. The causes of tastes and odors in drinking water. Water and Sewage Works, 104:6; 243, 1957.

4. Hale, F. E. The use of copper sulfate in control of microscopic orgaisms. Phelps Dodge Refining Corporation (ed. 1954).

5. Humphrey, S. B., and M. A. Eikleberry. Taste and odor control using $KMnO_4$. Water and Sewage Works, 109; Reference Number (October 1962.

Chapter 16

1. Caldwell, D. H., and W. B. Lawrence. Solution of water softening and water conditioning problems by chemical equilibrium methods. American Chemical Society, St. Louis, Missouri, September 7, 1948.

Recommended Additional Reading

Linsley, R. K., M. A. Kohler, and J. L. H. Paulhus. Hydrology for engineers. McGraw-Hill Book Co., New York, N. Y., 1955.

Fair, G. M., J. C. Geyer, and D. A. Okun. Water and wastewater engineering. Vols. 1 and 2, John Wiley and Sons, New York, N. Y., 1966 and 1968.

Babbitt, H. E., J. J. Doland, and J. C. Cleasby. Water supply engineering. 6th ed., McGraw-Hill Book Co., New York, N.Y., 1962.

American Water Works Association. Water quality and treatment. 3rd ed., McGraw-Hill Book Co., New York, N.Y., 1971.

Rich, L. G. Unit operations of sanitary engineering. John Wiley and Sons, New York, N.Y., 1961.

Weber, W. J. Physicochemical processes for water quality control. Wiley-Interscience, New York, N.Y., 1972.

Degremont. Water treatment handbook. 4th ed., Taylor and Carlisle, New York, N.Y., 1973.

Powell, S. T. Water conditioning for industry. McGraw-Hill Book Co., New York, N.Y., 1954.

Nordell, E. Water treatment for industrial and other uses. Reinhold Book Corp., New York, N.Y., 1961.

Rich, L. G. Environmental systems engineering. McGraw-Hill Book Co., New York, N.Y., 1973.

Salvato, J. A. Environmental engineering and sanitation. 2nd ed., Wiley-Interscience, New York, N.Y., 1972.

Campbell, M. D., and J. H. Lehr. Water well technology. McGraw-Hill Book Co., New York, N.Y., 1973.

Todd, D. K. Groundwater hydrology. John Wiley and Sons, New York, N.Y., 1959.

Walker, R. Pump selection. Ann Arbor Science Publishers, Inc., Ann Arbor, Michigan, 1972.

Appendix

Viscosity of Water at Various Temperatures[a]

Temperature °C	Density gm/cm^3	Kinematic Viscosity[b] Centistoke (10^{-2} cm^2/sec)	Dynamic Viscosity[c] Centipoise (10^{-2} gm/cm-sec)
0	0.99984	1.787	1.787
5	0.99996	1.519	1.519
10	0.99970	1.307	1.307
15	0.99910	1.140	1.139
20	0.99820	1.004	1.002
25	0.99704	0.8930	0.8904
30	0.99594	0.8008	0.7975

[a]Source: Handbook of Chemistry and Physics, 1970-71, The Chemical Rubber Company, Cleveland, Ohio.

[b]To convert to English Units multiply by 2.088 x 10^{-5}.

[c]To convert to English Units multiply by 1.075 x 10^{-3}.

CONVERSION FACTORS

1 ft = 0.305 m
1 ft^2 = 929 cm^2
1 ft^3 = 28.32 liters = 0.02832 m^3 = 7.4805 U.S. gallons
1 U.S. gallon (liquid) = 3.7854 liters
1 pound (mass)/ft^2 = 4.88243 kg/m^2
1 pound (mass)/ft^3 = 16.0185 kg/m^3
1 kg (force) = 9.806 N
1 kg (force)/cm^2 = 9.8 x 10^4 N/m^2
1 pound (force)/ft^2 = 47.8803 N/m^2
1 pound (force)/in^2 = 6895 N/m^2 = 705 kg/m^2
1 mm Hg = 133.322 N/m^2
1 atmosphere = 14.7 pound (force)/in^2 = 1 kg/cm^2
1 ft H$_2$O = 2989.07 N/m^2
1 metric horsepower (PS) = 735.499 Watts
1 hp (horsepower) = 1.014 PS = 746 Watts
1 kw = 1.36 PS
1 kg/ms = 1 Ns/m^2
1 m^2/s = 10^4 cm^2/s
1 rev/min = 0.10472 rad/s
1 Million U.S. gal/day (mgd) = 1.55 ft^3/s = 2.63 m^3/min
1 U.S. gal/ft^2/min = 2.45 m^3/m^2/hr = 2.45 m/hr
1 Million U.S. gal/acre/day = 0.937 m^3/m^2/d
1 U.S. gal/ft/day = 12.4 ℓ/m/d
1 ppm = 1 mg/ℓ = 1 g/m^3

Billion = 10^9
Trillion = 10^{12}

Index